ENVIRONMENTAL MUTAGENESIS and EPIDEMICS

SECOND EDITION

Globalization and Ecophagy

ENVIRONMENTAL MUTAGENESIS and EPIDEMICS

C.B.S.R. SHARMA

Pondicherry University

P.C. ANIL

Consultant - Forest Carbon Offsets -Origination and Validation VCSA Approved AFOLU Methodology Expert

Tampa, Florida, USA

GenNext Publication

ISBN: 978-93-5663-064-2 (PB)

First Published, 2023

Published by

GenNext Publication
5, Ansari Road
Daryaganj, New Delhi-110002
Phone: 9811692060
E-mail: gennextpublication@hotmail.com

Printed at: Gyan Press, Delhi.

DEDICATION

Dr. Gopal - Iyengar and Dr. K. Sundaram

Chiefs of Biology and Agriculture Division
and of Biomedical Groups, BARC,
Atomic Energy Commission, GOI. Bombay.

As founders of Environmental Mutagen Society of India in Baroda in 1975, along with Drs. Notani, Bhattacharya, Bhatia, Raghu, Aiyar, Murty, Chauhan and their deputies provided excellent leadership to scientists all over India in this field during the initial decades.

H.J. Miller

Discovered that Mutations can be produced by X-rays

Charlotte Auerbach

Discovered that Mutations can be produced by chemicals

Laurie Garrett

Conducted great inquiries on Epidemics

The present authors fear that Environmental Mutagenesis may become the mother of many future diseases.

Mutations - major and minor, recombinants in pathogens and harmless ones, may be involved.

ACKNOWLEDGMENTS

My pursuits in the field of Environmental Mutagenesis have been long and arduous punctuated by failures. So many people have helped me to sustain interest in this field of constant and continuous danger to mankind and ecosystems. I endeavor to recollect as much of these people as possible.

At B.I.T.S. Pilani, Love and Godliness was shown to all by Prof. B N Muley, the spirit of hard work and fortitude came from Dr. S.K. Pillai and Dr. B.D. Deshpande and the spirit of enquiry and analysis, from Dr. Gundurao – all my teachers. Vinca alkaloids were brought by Prof. Santra, of Pharmacy from USA. At Delhi, it was the lab and junior colleagues of Dr. M.S. Swaminathan, for everything in those terrible days.

At I.I.Sc, Bangalore, this field was introduced by Dr. R. Nagaraja Rao; interests sustained by Profs. S A Kumar and Sharat Chandra and a student, Dr. S S V Prasad, now at Houston.

At Berhampur two VCs, Justice Abhimanyu Misra and Justice BK Patra, HOD of Botany, Prof. H. Patnaik, Engineers Ram Mohapatra and Sarat Misra, and my Doctoral students – R K Sahu, B N Behera, S K Dash and B B Panda and their friends helped me to stabilize and progress for mutual benefits. At Waltair, Profs. BGS Rao, Bhairava Murty,

Seshavataram and their students – DSS Raju, Subba Rao and Vasudeva Murty helped me in the field study around the then Hindustan Polymers Factory, 40 years ago.

At Madurai my work could not have been possible but for the help rendered by Fr. Peter of Sacred Heart College, Shenbaganur and the dogged perseverance of my then students- Panneerselvam and Arumikkili. At Coimbatore the students were enthusiastic. Prof. C R Krishna Murty, then Director of ITRC, Lucknow evinced much interest in our field-cum-lab works.

At Pondicherry University by, I was not as lucky as in the other places - with Authorities and Colleagues, educated Indians, greatly deteriorating due to globalization. This is evidenced by the cadaverous treatment I got from most around me.

I have gone through many turbulent times. Here comes the value of the family which always stood by me – my wife, Ramadevi, my daughter, Swati, son-in-law Dhanraj and grand sons, Aaditya and Arjun. If my career in this field is maturing into this book, credit entirely goes to them for giving me peace and tranquility.

DEFINITIONS AND COMMON SENSES

ECOLOGY Inter- relationships and interactions among soil, water, air, plants, animals, microbes and to humans establish a balance and maintain it with climate and others, if any. This evolutionary conscience grew during the last 50 years although developing for more than 100 years

GLOBALIZATION All pervading licences to all the mighty with no inhibitions across all the existing borders. Decimation of Human Spirits, Guerilla-Communists, Red-shirt Communists, White- Shirt Communists, Righteous People of any hue, innocents and the simple

ECOPHAGY Age-old Ecosystems are simply swallowed, interlinks and balances vanishing, never to return. To correct errors here ecosystems elsewhere will be destroyed

GENE Determinant of inheritance or heredity in all organisms, the DNA –Deoxyribose Nucleic Acid

MUTAGENS Y-Rays and a myriad Chemicals altering DNA permanently.

MUTATION The process of DNA alteration, which are of many types and routes

MUTANT The genetically altered organism, which will be different from the parental, the wild one

MUTAGEN BURST Forceful Emission of gaseous or liquid pollutants into the ambient environment, which may be mutagenic

EPIDEMICS Decease caused by mutual contacts of nay sort, therefore contagious, fast spreading and difficult to control. If it crosses boundaries of any type, it is called pandemic. If its origin is genetic mutation, it is Gene-Epidemic, the most difficult to manage or control, even to understand

DIASPORA People of one nationality living / domiciling in another nation through migrations, of any kind. In India People of one language or state living in another state for historical reasons are generally despised despite their correct routes, services and uses – generally 2nd class citizens and if unfortunate 3rd class – like the present author!

ASPIDISTRA A Lily of eternal hope in times of strife, especially in Britain during wartime, made famous by Orwell. Lord Krishna of Mahabharata was known to wear a garland of these CHENGALVA flowers. One of the largest distributed plants of the world, its flowers are simple and elegant, long-standing, leaves ever-green, untouched by fungus, insects and microbes.

Only Man Does!

PREFACE

FIRST EDITION, 2020

This book is a product of righteous indignation at some recent happenings in my country: 1. Covid 2, Styrene Burst in Vizag and aspects associated with them.

A. **Covid-19:** The conduct of Indian states towards the migrant labour helping them in development. They were brought from other states by agents and employed by contractors in projects allotted by the governments. When the going got bad, they were abandoned and left for themselves everywhere. They formed an exodus of about a Crore of people, as if during partition times. Where would they go in those curfew times? This immoral, sinful performance was well documented. All governments and political parties and the administration are fully culpable for this tragedy.

   The migrant workers, on their part, should have rebelled, but submitted, as if it is their fate! They have to demand, if not loot, or rob and remain stay put. They have lost that spirit of justice and will to fight for fair play, like most others in India

   Culpability for Covid Tragedy is with all governments became complacent with scientific rescarch specially pertaining to Microbes, after the globalization. They also become ignorant with

Coming Plagues and other warnings, They were ignored in the euphoria of openness when technology was mistaken to be science and critical science as pessimism or skepticism. Every one appear to have a little culpability.

Covid is a Mutant, caused by Mutagens by a mechanism of Mutagenesis.

B. **Styrene Burst in Vizag:** In this accident in a polymer factory, Styrene chemical escaped from containers to everywhere far and near – not a spill but a flood and berserk. Everything here was shameful – falsehood, escapism, secrecy, misleading, bribery, Ignorance, negative politics – by one and all. The CM, Government, Police, administration, media, Health Department, Scientists – all indulged in cheating people – nay –indulging in self deception!

My interest in anything bad for the country stems from my patriotism. This increases when some facts are known to me revealing others follies. When they appear to be either intentional or due to ignorance, they appear unforgivable.

Concerning Styrene so much is known. Yet nothing is visible. The matter was cunningly suppressed, giving an impression that everything is managed and all is well. Facts are anything but this. I have interest in this scenario fro mid-70s, when I was working in Berhampur and looking for polluted ecosystems. I happened to survey this place and even initiate the first studies. For some reasons this could not be continued. This is one of the worst scandalous management to silence citizenry,

Styrene is a Mutagen whose spread, impacts and ecology are unknown.

C. **Mutations and Gene-Epidemics:** Covid is a mutant; caused a gene-epidemic, the causes are not known. Styrene is a mutagen, but its targets and effects are not known. Therefore an understanding of Mutagenesis and of Gene-Epidemics complete the stories of both tragedies. Consequently the first part is devoted to Mutagenesis and Gene-Epidemics, the second to Styrene burst and the third to Covid epidemic, No one could have imagined this queer association between the virus and the chemical!
Science is like this!

D. **Mutations ignored by Greens, COR, and UNCED and Scientific Bodies: The** possibility of Epidemics from Genetic origins was not envisaged by the above organizations although they have covered all the possible Ecospasms. Even the Mutagen societies did not have the imagination to weigh the ramifications of mutation process. Furthrt they lacked good public relations to influence national policies. Strangely it was left for a Journalist to suspect such a possibility, but ignored by the earlier mentioned watch dogs of human welfare.

All these were some historical errors by psyches conditioned by the euphoria generated and sustained by licentious governance called in India, Globalization. A few benefits often blur visions and mask futures. Unfortunate!

E. **End of Civilization May begin with Gene-Epidemics:** There is a considerable body of opinion that the beginning of our civilization has begun. This kind of future was envisaged by some of the votaries of LTG world view of COR advanced as back as 1972. But the same authors, years later prophesied that this can be managed with sufficient efforts in several human activities. This was criticized for its Malthusian content. Interestingly the doomists predicted the collapse of civilization by the middle of this century, starting from the third decade ! The issue is about the symptoms.

Climate Change is prominent in this scenario. Forest fires, Rains and Floods, Crop failures, Human Migrations – in short all disasters, But I consider the damages from the Microbes as the most telling. Garrett was suggestive of this in 1996, and Covid is here. I envisaged this possibility in 1985 in agro-ecosystems, extended it in 2002 to ecosystems and developed the theme by 2009 when we predicted Gene - based epidemics which could become Pandemics.

Basis for these are data and reasoning about possibilities. In fact the situation about mutations is that many chemicals in small doses and complex mixtures pervading the ecosystems can cause mutations in any organism through a variety of exposures.

To a positive mind all problems are surmountable by human intelligence and endeavors but for the politics among the rulers and among the ruled. The former is afflicted by immoralities currently experienced prominently in the state of AP, soon to spread out. The latter pervaded through out India during globalization of 30 years and recently and more vigorously since march this year. When psyches are altered, cultures vanish, political suicides become unpreventable. This book in three parts, is meant to vent out my forebodings. Therefore it may be personal, anecdotal but scientific as well as experiential. Hope there will be some souls to share my pains and views, provided providence grants me the time to complete this venture!

SECOND EDITION, 2022

This is entirely due to criticism and advices of friends. A few photographs could not be obtained. It is believed that most of the new diseases are epidemics in nature and originate due to silent nature of the mutations. And mutations can occur in the environment ubiquitously.

CONTENTS

1

INTRODUCTION

In this year three surprises shook India, among others. They appear unrelated. But to me they are:

Firstly, Covid-19, is a mutant. Mutants are generally deleterious but recessive and mutate repeatedly and loose virulence. But not this one! Secondly, Styrene that is emitted profusely and spread in Visakhaptatnam is a mutagen that can cause mutants. Thus these two are the two sides of a coin. But people did not realize. My intention is to emphasize this fact of abundant mutagens or their precursors as well as prevalence of mutants but not pathogenic enough in various ecosystems.

The third is the human conduct which was witnessed by the whole country, towards the migrant labour and their meek submissions.

Our Values, Concerns, Psyches and Cultures have metamorphosed due to the impacts of Globalization which mutated Indians – get –rich-quick by any means and consume what the modern technologies give us. Consequently we loose sight of our duties, vigilances, ideas, what not – if every thing comes free with a price why to work or think. Everything happening in India, including politics, can be explained from this stand point. For both the first two I blame the third.

I wish to bring together the mutations, causing Covids on one hand, and chemicals like Styrene which produce mutations which are related by the human actions.

Corona is a non-living particle containing RNA and Protein, called a virus, getting activated to multiply in a living system. It gets mutated and transmitted to humans; or transmitted first and mutated; It gradually shows harmful effects on the host becoming a pathogen. But why is this mutation - directly by a radiation or a chemical, or by a mixture of chemicals in nature, Pollution ? Thus the story began in China a few months ago and spread through out the world.

A. CORONA

Corona is an example for a genetically changed virus, which, unfortunately became pathogenic to humans and infective horizontally. But an important question is the causes behind - the whole subject of Genetic Toxicology or Environmental Mutagenesis. Chemicals in air, water, soil, usage, food and food chain, normally in minute quantities and in mixtures bring about drastic changes in the genetic material rendering it harmful to humans, i.e. named as COVID-19 by the WHO.

B. MUTATIONS

Mutations occur sporadically in all micro-organisms of all habitats. They are normally recessive and reversible, not affecting the humans adversely. Some pathogens become worse or better, affecting human societies. Some repeatedly mutate. These phenomenon have been witnessed by the recent societies in resistances to drugs useful until then. Drugs themselves induce mutations in the pathogens to resist them, familiarity, probably breeding contempts! *e.g. Tuberculosis, Malaria, Influenza, etc:*

1. Some only become pathogenic, otherwise innocuous. That these can extend themselves from agricultural field to food science and medicine and cause immeasurable damage to human societies has been feared from nearly 30 years by several scientists including this author.
2. This is mainly because of occurrence of birth defects, heritable deceases, and cancers in human societies. On the other hand, use and misuse of chemicals in all walks of life, pollution of air, water and others by complex mixtures often in small and immeasurable quantities, occurrence of similar defects in animals of many types in nature, have been long felt.

   Suffice it to say this is an unexpected extension of the pathogenic mutations.
3. Mutation theory that goes back to 12 decades to De Vries and 60 years to the heydays of Mutation Breeding to improve crops. Mutations contribute to variations in nature and add to evolutionary forces.

   They cause changes in the agro-ecosystems which could be both beneficial and deleterious to human health both directly and indirectly causing resistances in pathogenic microbes and their vectors.

C. EPIDEMIC DECEASE

This is infectious but not contagious. So can be an epidemic, and is, Socially, as people move, it tends to become pandemic. This can be controlled. Then only the epidemic can be contained and researched into how an innocuous virus suddenly become disastrous. If this is not done similar one can happen any where and spread. A warning has come that gene-epidemics can become pandemics.

D. GENE- EPIDEMICS

Thus we have before us two questions. First is the origin of mutations and the spread of the mutants. Second is the latitude of effects, The later was limited first only to humans and to mutagenicity, teratogenecicity and carcinogenicity. This was soon found inadequate leading to the theory of Gene- Epidemics. These three nearly irrecoverable changes, resistances of pests and pathogens in agri-, aqua, and veterinary science; Pest resistance by vectors, human infections becoming resistant to drugs and finally even the origin of new pathogens - all happen from mutations. These effect human society. Therefore the impacts are named as Gene- Epidemics and the sources of mutations were already known as mutagens. And the process, Mutagenesis. At hand we have a whole new avatars of old figures needing to be managed and maneuvered using all intelligence at the disposal of the human species. Or Else ?

E. STYRENE BURST AT VISAKHAPATNAM

The disaster in the former Hindustan Polymer Factory at Vizag from where floods of Styrene Gas was emitted into the environment is by any means a shameful hush-up. First the chemicals is a genotoxin and killed several people and hurt many others. They were all shut and others silenced. What is needed is an Ecosystem Approach first to find the magnitude of accident and later the ramifications, which alone can yield a map of boundaries to investigate. The longitudinal approach involves monitoring the ingalers for some generatations, assuming that they counted the victims correctly. Water and agri fields too needed a deeper probe. But none were done and I doubt whether the state of AP has the facilities or the expertise or the awareness of what is infront them and in store!

F. CIVILIZATION

An increasing body of intellectuals has been fearing the possibility of an end to the present civilization. They advance reasons such as climatic changes, rainfalls, floods, forest fires etc and the Malthusian and the Limits to Growth Fears. As a teacher of Environmental Sciences I too am a votary to this. I add the Gene-Epidemics arising out of mutations to this. Only this can be far ominous and dangerous as the world witnessed for the last 10 months.

2

HISTORICAL PERSPECTIVES

REDEEMING EARTH, ENVIRONMENT AND BIOTA

FORCES AND ACTIVITIES

50 years after the original events began to save the earth for the present and posterity, we are still at the same condition, probably worse. But efforts cannot slacken; history needs to be recollected, however distasteful it is for most of us, especially when there is an opinion that the beginning of end has already taken place or beginning. When I shifted from Botany at Berhampur University to Environment at Madurai University in 1980, I am to teach subjects I did not know and each begins with the basics and then history. The literature was so fascinating, new and frightening as well as hopeful.

During my 5-Yr stay and teaching at Madurai University, I Improved continuously and shared with students.- who come from different fields in science. A whole new world was opened, they felt – Sometimes the classes were held under trees, sometimes in the canteen. Often students brought me inputs to be shared by all. It was a glorious time. My interest and new knowledge was so impressive at Coimbatore that I could initiate a New Department of Environmental Sciences, and move within 2 years to Pondicherry Central University. There was a strange situation here. Students are young, excellent and full of glow and warmth, but the teachers recruited are of rock bottom standards. Most had neither knowledge nor experience. They used to spy on my lectures, poke fun at the mention of Club of Rome or talk of Reds and Whites and so on and try to spoil students by a myriad ways – religion, caste, language – how to alienate. This is the destiny of Environment in the hands of evils masquerading as teachers

A few years later, several Christian students requested me repeatedly for a write-up of the historical aspects on Environmental Dilemmas, which I did, probably in early '90s. The present one is an updated version of a small part of it! Righteousness and honesty accompanying sincere work did not interest people –economic globalization and consumerism everywhere. Colleagues failed me but they did not improve! This is the fate of environmental causes in so many places in the world and of many in India. And redemptions and recoveries are rendered alien. Alas!

INTRODUCTION

Recent times are beset with issues and problems of environment all over the world. This future was feared several scientist of the planet from as back as post second world war. But they went to deaf years ! It took 2/3 decades to realize the deteriorating environment. Thus not only the concerned and conscientious citizens but also elite elements realized the urgency of deterrent actions. Soon governments too realized as the disasters in environment have been increasing both in space and intensity.

Thus there were three major efforts concerning the whole globe which interestingly and significantly continue even now. The Conscientious people became THE GREENS, Die Grunen of Germany, who practiced green politics ever since 1972. The elite formed a "Club of Rome" and involved in intellectual activities of global relevance. With these around the governments could not remain inactive. With the initiative of Sweden United Nations initiated the periodical

"Earth Summits". The first meet took place at Stockholm – United Nations Conference on Human Environment [**UNCHE**] . More details of all these are given briefly for the advantage of the uninitiated.

All these people have more or less covered all issues – basics, problems, solutions and what not, but this subject that lead to Covid-19, The Mutagens. However, a small group in USA focused their energies only on this area almost at the same time. More details of this too is given although a miniscule outfit in comparison of the three others. This exclusively deals with Mutagens, whose activity resulted in the present Covid-19 misery, rising the possibility of more to come. This terrible possibility, strangely, eluded not only the thoughts of the three global forces but also went beyond the ken of imagination of the EMS of USA and IAEMS.

A. CLUB OF ROME

The Club of Rome was founded in April 1968 by Aurelio Peccei, an Italian industrialist and Alexander King, a Scottish Scientist. It was formed when a small international group of people from the fields of academia, civil society, diplomacy and industry met in Rome, hence the name.

1. **The *Problém***
   (a) Central to the formation of the club was Peccei's concept of the *problematic*. It was his opinion that the problems of mankind-environmental deterioration, poverty, endemic ill-health, urban blight, criminaliy - all are interrelated. "It is this generalized meta-system of problems which we have called the "problematic"

(b) In 1970, Peccei's vision was laid out in a document written by Hasan Özbekhan, Erich Jantsch, and Alexander Christakis. Entitled, *The Predicament of Mankind; Quest for Structured Responses to Growing Worldwide Complexities and Uncertainties: A PROPOSAL.*

2. ***The Limits to Growth***

(a) The Club of Rome stimulated public attention with its first report, *The Limits to Growth.* Published in 1972, its computer simulations suggested that economic growth could not be indefinite because of resource depletion. The 1973 oil crisis increased public concern further. The report sold 30 million copies in over 30 languages, making it the best-selling environmental book in history. *Mankind at the Turning Point* was the "second report" to the Club of Rome in 1974 authored by Eduard Pestel and Mihajlo Mesarovic. In addition to providing a regional breakdown, they gave a more optimistic prognosis for the future of the environment, noting that environmental and economic catastrophe were preventable.

(b) In 1991, the club published *The First Global Revolution.* That pollution, the threat of global warming, water shortages, famine are the enemies of all. In their totality and their interactions these phenomena do constitute a common threat which must be confronted by everyone together. But, we fall into the trap of mistaking symptoms for causes. All these dangers are caused by human intervention in natural processes, and it is only through changed attitudes and behaviour that they can be overcome. The real enemy then is humanity itself."

(c) In 2001 the Club of Rome established a think tank, called *tt30*, consisting of about 30 men and women, ages 25–35. It aimed to identify and solve problems in the world, from the perspective of youth.

3. **Organization**

The Club of Rome is a membership organization and has different membership categories. Aside from full members there are associate members, who participate in research and projects.. The club also has honorary members, some of whom are: Manmohan Singh, Mikhail Gorbachev, Václav Havel, Ashok Khosla, Pierre Elliott Trudeau, Nagendra Singh, The annual general assembly of 2016 took place in Berlin in November. Among the guest speakers were former German President Christian Wulff, German Minister Gerd Müller, and Nobel Laureate Muhammad Yunus.

4. **National Associations**

The Club has national associations in 35 countries and territories. The mission of the national associations is to spread the ideas and vision in their respective countries, to offer solutions and to lobby for a more sustainable and just economy in their nations, and to support the international secretariat of the Club with the organization of events, such as the annual general assembly.

5. **Current Activities**

   By 2017, 43 reports were made by the COR. The most recent is *Come on ! Capitalism, Short-termism, Population and the Destruction of the Planet*. In 2016, the club initiated a new youth project called “Reclaim Economics”. With this project they support students, activists, intellectuals, artists, video-makers, teachers, professors and others to “shift the teaching of economics away from the mathematical pseudo-science it has become.”

6. **Critics**

   Nobel prize-winning economist Robert Solow criticized *The Limits to Growth* as having “simplistic” scenarios. He has also been a vocal critic of the Club of Rome, ostensibly for amateurism. Mathematicians Vermeulen and De Jongh has shown “very sensitive variations” and having “dubious assumptions and approximations”. They claim that the Meadows et al. methods, data and predictions are faulty, that their world models and their Malthusian bias do not accurately reflect reality.

B. UNITED NATIONS CONFERENCE ON HUMAN ENVIRONMENT

Large-scale industrial pollution, the growing threat of nuclear radiation, mass destruction of ecosystems around the globe, the advent of international concern about a global environmental crisis heightened global awareness.

It was the initiative of Sweden that laid the foundation for international cooperation.

In 1967, UNO initiated concerns to "facilitate co-ordination and to focus the interest of member countries on the extremely complex problems related to the human environment." It was the Swedish delegation, led by Sweden's UN Representatives. who took the initiative to convene the first United Nations Conference on the Human Environment (UNCHE) in Stockholm in 1972. By the late 1960s, Swedish diplomats used the UN system to shift the diplomatic focus towards environmental protection.

> Two years after their initial proposal, on 15 December 1969, UN resolved that: "It should be the main purpose of the Conference to serve as a practical means to encourage, and to provide guidelines for, action by Governments and international organizations designed to protect and improve the human environment and to remedy and prevent its impairment, by means of international cooperation, bearing in mind the particular importance of enabling the developing countries to forestall the occurrence of such problems".

On 5 June 1972, the UN Conference on the Human Environment opened its first plenary session in Stockholm. Swedish Prime Minister Olof Palme criticized the industrialized world for its ecological and economic exploitation, causing the world's greatest environmental problems at the expense of developing countries the

114 Governments reached an unprecedented level of agreement mostly due to the efforts of Maurice Strong, UNCHE not only established a remarkable Declaration of Principles, but also came up with the necessary institutional arrangements for international cooperation in environmental protection, the creation of the UNEP.

Stockholm Declaration

The meeting agreed upon a Declaration containing 26 principles concerning the environment and development; an Action Plan with 109 recommen-dations, and a Resolution. Principles of the Stockholm Declaration:

1. *Human rights must be asserted, apartheid and colonialism condemned*
2. *Natural resources must be safeguarded*
3. *The Earth's capacity to produce renewable resources must be maintained*
4. *Wildlife must be safeguarded*
5. *Non-renewable resources must be shared and not exhausted*
6. *Pollution must not exceed the environment's capacity to clean itself*
7. *Damaging oceanic pollution must be prevented*
8. *Development is needed to improve the environment*
9. *Developing countries therefore need assistance*
10. *Developing countries need reasonable prices for exports to carry out environmental management*
11. *Environment policy must not hamper development*
12. *Developing countries need money to develop environmental safeguards*
13. *Integrated development planning is needed*

14. *Rational planning should resolve conflicts between environment and development*
15. *Human settlements must be planned to eliminate environmental problems*
16. *Governments should plan their own appropriate population policies*
17. *National institutions must plan development of states' natural resources*
18. *Science and technology must be used to improve the environment*
19. *Environmental education is essential*
20. *Environmental research must be promoted, particularly in developing countries*
21. *States may exploit their resources as they wish but must not endanger others*
22. *Compensation is due to states thus endangered*
23. *Each nation must establish its own standards*
24. *There must be cooperation on international issues*
25. *International organizations should help to improve the environment*
26. *Weapons of mass destruction must be eliminated*

One of the seminal issues that emerged from the conference is the recognition for poverty alleviation for protecting the environment. the connection between ecological management and poverty alleviation.

The Stockholm Conference motivated countries around the world to monitor environmental conditions as well as to create environmental ministries and agencies. Despite these institutional accomplishments, including the establishment of UNEP, the failure to implement most of its action programme has prompted the UN to have follow-up conferences.

Some argue that this conference, and more importantly the scientific conferences preceding it, had a real impact on the environmental policies of the European Community). For example, in 1973, the EU created the Environmental and Consumer Protection Directorate, and composed the first Environmental Action Program. which has led to the Kyoto Protocol and the Paris Agreement, and has given a foundation of modern environmentalism.

C. DIE GRUNEN AND GREEN POLITICS

1. Introduction

Unease about adverse consequences of human actions on nature predates the modern environmentalism. Social commentators from ancient Rome and China complained of air, water and noise pollution. The modern roots of environmentalism can be traced to thinkers such as Rousseau, Theoreu, Mohandas Gandhi, Spinoza and the values of indigenous peoples,

Green movement has emerged in the environmental crisis - a series of interconnected crises caused by population growth, air and water pollution, the destruction of the tropical and temperate rain forests, the rapid extinction of entire species of plants and animals, the greenhouse effect, acid rain, ozone depletion, etc, the ecospasms. Many are by-products of technological innovations, but are also broadly cultural and political.

Such destruction is a consequence not only of large-scale policies but of small-scale, everyday acts. Therefore the adage: "Think globally and act locally."

(a) An environmental or ecological or land or Planetary ethics, would emphasize the web of interconnections and mutual dependence within which we and other species live. Adherents to green politics consider it to be part of a higher worldview

(b) Organised environmentalism began in late 19th Century Europe and the United States as a Reaction to the Industrial Revolution with its emphasis on unbridled economic expansion. "Green politics" first began as conservation movements; for example the Sierra Club, founded in San Francisco in 1892.

(c) These people conscientious of human environment draw terminology from deep ecology, feminism, pacifism, anarchism, socialism, democracy, eco-socialism, and social ecology. Further they share many ideas with the ecology, conservation, environmental, feminist and peace movements. In addition to democracy and ecological issues, green politics is concerned with civil liberties, social justice, nonviolence and participatory democracy. In the 1970s, Green politics arose as a new philosophy which synthesized their goals

2. Cardinal Principles

The German Greens drafted the earliest statement on environment, the Ten Pillars, which have been followed by many green parties worldwide as a foundational statement of the green ideology: These Ten were limited to six in 2001 by the Global Greens Charter. They are:

> Ecological wisdom, Social justice, Grassroots democracy, Non-violence Decentralization, Community-based economics, Post-patriarchal values [Feminism] Respect for diversity, Global responsibility, Future focus

As such, the ideal Green Party is thought to grow from the bottom up, from neighborhood to municipal to (eco-) regional to national levels.

3. **Political History**
   (a) In June 1970, a Dutch group called Kabouters with Groene Plannen won 12 of the 45 seats in 6 city councils. The first political party to be created with its basis in environmental issues was the United Tasmania Group, *Values Party*, founded in Australia in March 1972. This inspired the creation of Green parties all over the world. Europe's first green party, PEOPLE in the UK came into existence.
   (b) The German Greens, contended in their first national election in the 1980 federal election. felt their interests were not expressed by the conventional parties. They put forward their Four principles which were accepted by many Green Parties around the world. The term "Green" was coined by one of the founders of the German Green Party, Petra Kelly. In the 1983 federal election, the Greens won 27 seats in the Bundestag.

They adopted the sunflower symbol. Critiques called it a watermelon!

4. **Grades of Green**
   (a) Earth is looked upon the as a benevolent deity-the goddess Gaia to be worshiped in reverence and awe. In this way we can liberate ourselves from the restrictive rationalism of the modern science. Other Greens disapprove of such a spiritual or religious orientation,
   (b) Other differences relate to political strategies and tactics. Greens take an active part in electoral politics, following the lead of Greens in Germany. Aware of the formidable obstacles facing minority parties, most have favor to work within existing mainstream parties (Democratic Party in the US], or hiring lobbyists to influence legislation. Greens of the "bioregionalist" persuasion feel that such political organization should be based on natural, rather than artificial or political, boundaries regions.
   (c) *Some groups, such as Greenpeace, Sea Shepherd Society and Earth First favor* **dramatic** direct action by publicizing and protesting practices destructive of the natural environment and capture public attention. Such militant tactics are decried by moderate groups, which favor efforts to influence legislation and inform the public on environmental matters. The Sierra Club, for example, lobbies Congress and legislatures to pass environmental legislation. It also publishes books and produces films and videos about a wide

variety of relevant al issues. Similar strategies are followed by other groups, such as the Environmental Defense Fund. Another group, The Nature Conservancy, solicits funds to buy land for nature preserves.

5. **Global Organization**

Global Green networking dates back to 1990. Following the Planetary Meeting of Greens in Rio de Janeiro, a Global Green Steering Committee initiated Global Green Network including a Global Green Calendar, Global Green Bulletin, and Global Green Directory. In 1996, 69 Green Parties from around the world signed a common declaration

Despite their political differences, Greens are alike in assuming that all things are connected. Ecology is, after all, the study of interconnections-and they agree that complex ecosystems and the myriad life-forms they sustain are valuable and worthy of protection by political and other means. However, the efforts of the Green Parties shows their potential to unite and campaign jointly.

D. IAEMS, IAPEMC, EMIC

International Association of Environmental Mutagen Societies
International Agency for Protection against Environmental Mutagens & Carcinogens
Environmental Mutagen Information Centre

Almost around 1970s, when the world was preoccupied and dominated by Die Grunen, UNCHE or the Club of Rome there are some scientists scattered in the developed world, who were concerned about the health of people in their societies, for, these are not the usual and common place maladies that will be any way taken care by the health systems. These ailments appeared then insoluble once set in.

Some of them also appeared to be similar to the ones caused by the radiations, as experienced by the Crow's Committee on biological effects of Nuclear Explosions in Japan. They are Birth defects such as Polydactily and Cleft lip, Cancers, and Heritable diseases such as Achondroplasia, albinism, Retinoblastoma. and phenylketonuria, These are christened as Mutagenesis. Carcinogenesis and Teratogenesis– The inducing agents, Mutagens, Carcinogens and Teratogens respectively. These were originally called Radiomimetics as they were thought to mimic ionizing radiation in their actions.

All the above ailments seem to increase yearly with no sight of recovery, Increasingly it was felt that a mutation process is a root cause and that this must happen first before onset of cancers or birth defects. Therefore all concerned scientists practicing biology, genetics, microbiology, toxicology, pharmacology, teratology, carcinogenesis, epidemiology, and public health, met and formed an association at ORNL, USA, Environmental Mutagen Society. In 1969. It has received wide support and soon could greatly increase its activities.

Thus similar societies and activities sprang up in UK, Germany, Japan, China, India, etc. They soon formed into a federation, International Association of Environmental Mutagen Societies, With in USA they formed an EMIC, Environmental Mutagen Information Centre which is closely connected with CASR – Chemical Abstract Service Registry.

All these led soon to the formation of ICPEMC, International Council for Protection against Environmental Mutagens and Carcinogens, in the lines of ICPR against Radiations. All these took about half a decade only. In India EMSI was formed in 1975.

EMSUSA conducted annual meetings, maintains a prestigious Journal, news letter, conducts conferences, international Conferences, triennially supported by industries increasingly to encourage participation from developing countries to encourage talents. I happen to be among the first list of awardees, some 35 years ago. Also I was the only one from India invited to the 3rd ICEM at Cleveland and given membership in their organizing Committee. I also served as a Referee to articles in the British Journal, Mutagenesis.

As an example from Developing societies, EMSI began functioning from 1975 entirely due to the scientists of BARC from the Divisions of BARC – biology and Agriculture and Biomedical groups under their leadership. These efforts continued for 20 years, until pseudo- scientists brought it down for their own reasons to spite some one. The three villains came from Pondicherry, Coimbatore and Madras. Gone are those days - Symposia, Work Shops, Edited Volumes, News letters, Annual Meets and an enviable status in the scientific circles In India and abroad. It is now a just society – a non-descript unknown entity, far removed from the illustrious BARC, typifying Indian science.

CRITIQUE

How come that all the intelligentia of the world represented by the three mainstreams of efforts and the exclusive societies of the scientists MISSED the pervading impacts of chemicals and possible disasters due to their interference with genetic material!!!

1. The three major forces working on people see reason to make efforts to let the common resources remain secure. They were striving hard for the last 5 decades. How come none among them gas doubted the possibility of chemicals to be dangerous, if not more than radiations. In fact it now appears so! Radiomimetic chemicals, laterly Mutagens did not smell familiar to them. Otherwise, we would not have a Covid today, for it is only a mutant. Therefore a mutagen must have been there, known or unknown.
2. On their part the mutagen societies did not bother to have good public relations. They ought to have given wide publicity. It is natural for scientists to shun publicities. But these substances pose enormous risks – This they did not foresee. Because excessive imagination or speculations are equally frowned.
3. But I did some speculation on gene-epidemics in 1985 to a leader-level international audience at Penang, Malaysia, but confined to agro-ecosystems, which could have been extended to human systems by some enterprising and imaginative scientist ! Again it was left for me to do this many years later. I also warned that mutagens may be unknown and flourishing in the ecosystems, to an audience of young professionals at MSSRF, Chennai. This supports the claim of Chinese that Covid - causing Corona came from a natural ecosystem. My final imagination ended a decade ago when I expanded the Theory of Gene –

Epidemics, connected it to Epidemics, and hinted at the possibility of Pandemics to a national audience brought by none other than ICMR at Bangalore in their National Institute for Occupational Health.

Scientists had to be imaginative and vigilant, for no one knows what comes from where and when. This is what not only the three major movements to protect humankind and specialist societies could not do, the august audiences I spoke to at three different places did not care. If these happened, I felt that Corona could not have become so threatening.

REFERENCES

1. *Mesarovic, Mihajlo; Pestel, Eduard (1975). Mankind at the Turning Point. Hutchinson.*
2. Alexander King & Bertrand Schneider. *The First Global Revolution*, 1993. p. 70.
3. Von Weizsaecker, Ernst; Wijkman, Anders (2018). Come On! Capitalism, Short-termism, Population and the Destruction of the Planet. New York: Springer.
4. Cole, HSD (1973) Models of doom; a critique of The limits to growth USA Universe Books.
5. Suter, K. (1999). "The Club of Rome: The Global Conscience". *Contemporary Review*, 275 (1602), 1-5, Donella Meadows Institute.
6. Reclaim Economics. Club of Rome, 2017.
7. Ivanova, Maria. "Designing the United Nations Environment Programme: A Story of Compromise and Confrontation." *International Environmental Agreements* 7 (2007): 337-61.
8. Strong, Maurice. "One year after Stockholm: An ecological approach to management." *Foreign Affairs* 51 (1973): 690-707.
9. Ward, Barbara & René Dubos.*Only One Earth:The Care and Maintenance of a Small planet.*
10. Capra, F., and C. Spritnak. *Green Politics.* New York: Dutton, 1984.
11. Manes, C. *Green Rage: Radical Environmentalism and the Unmaking of Civilization.* Boston: Little Brown, 1990.

12. Porritt, J. *Seeing Green: The Politics of Ecology Explained.* Oxford: Basil Blackwell, 1984.
13. Worster, D. *Nature's Economy: A History of Ecological Ideas.* Cambridge: Cambridge U. Press, 1977.
14. Pesticide induced gene-epidemics in the Third World countries. (in) Global Development and Environment Crisis - Has Humankind a future ? (ed. Idris): 642-662. 1988.
15. GENEPIDEMICS & GENEPANDEMICS IN THE ECO-SYSTEMS AND HUMAN SOCIETIES* An Ominous Possibility emerging from a Reconsideration of Darwinism in its 150th year * A Trans-Disciplinary Approach - Environment and Health - Beyond. ROCHS-NIOH-ICMR- Complex, Bangalore, 16-19, September, 2009.

3

CHEMICALS AFFECT PLANT CELLS

I received the doctorate degree from BITS Pilani in March 1969. I was than working as lecturer in botany in DAV College, Muzaffarnagar, UP. In response to an advertise ment for the post of lecturer in cytogenetics at Indian Institute of science at Bangalore, I was interviewed and it was a great success.

I was selected but given the post of senior research assistant because I did not have post doctoral experience in developed country with advance implements. My pay was equivalent that of lecturer at the starting point, I was told by the HOD to upgrade Cytogenetics towards Bio-chemistry. This was in November 1969.

We have a habit of visiting a near by canteen in forenoon and afternoon. Habituated to gossip of course an academics and science most customers stay they for long. Most faculty and juniors see each others without talking with each other. In one such gossip sessions, a stranger approached our table and asked me whether I am interested in cells. Yes was my answer adding that responses of cells to stresses has been my interest, he came next day and sat with us he suggested, why not I review this subject has no such attempt was made. I agreed. An association began. He was Dr. S.A. Kumar a faculty from bio-chemistry.

My attempt began with cells. After a week it become limited to plant cells, after another week it further reduce to meristematic cells, by the end of the month I settled with plant, root, meristems. So much was the literature; only we did not realize, for, we never survey. He advised me to always review your work to realize where you stand both in thought and actions in your field of science. Dr. Kumar regularly took part to the progress of work and pattern of thinking he told later he was afraid of middle drops so common against scientists. Nevertheless, he was a good card player!

It took nearly 6 months to make the first draft. Regularly working in the library during evenings attracted the attention of Prof. Sathish Dhawan the director and some other senior professors. Biological abstracts bunches of papers with notes will always be at one particular place with a note "Please do not disturb"

The script was read by Professor Kumar, Dr. Sridhar and Dr. Viswanath of IISC and Dr. Kasthuribai of Bangalore University, it took another month taking all criticism into consideration and readability, the most. Professor Kumar forwarded it to CSIR journals. Dr. P.S. Jhangi accepted it for publication in Journal of Scientific and Industrial Research. He appreciates my effort and passed further similar sentiments from the reviewers from Delhi University and Baba Atomic Research Centre at Bombay.

The department office was amused as to why I was researcing may reprint requests. The HOD called me to enquire when I told him he was so happy to comment on my future in science that I was doing the right thing, alone yet 27 plus.

The review was published. I got 75 reprints. There were more reprint requests. Our office satisfied most of them I received personal letters appreciating me. Many thought I am a senior person just return from abroad. At least a score of them exclaiming when they knew. I was the author so young. Many of them become my long term friends and well wishers. Jealousies and prejudices are uncommon then, if at all existed.

REFERENCE: *Sharma,C.B.S.R, 1971. Effects of Chemicals on Mitotic Plant Cells. J.Sci. Ind. Res., 30: 571-587.*

Ever since the pioneering studies of Berenblum and others, work on the effects of chemicals on cell division has been mainly motivated by: (i) a search for means of inducing variations in the genetic material, (ii) the desirability of understanding the physiology of mitosis, and (iii) the importance of screening chemicals for anti-neoplastic activity. The exact classification of the chemicals involved or of the effects produced in the dividing cells must normally be based on the biochemical mechanism of actions of these chemicals and their correlation with the visible effects in the cell. In the absence of such information, the only possibility left is to merely catalogue the visible effects.

Many reviews on the antimitotic effects, dealing with cytogenetical and/or pharmacological aspects, have appeared, but most of them, barring a few, are either restricted in scope or did not aim to differentiate between plant and animal cell responses to antimitotic substances. There is no review so far which deals exclusively with the effects on plant cells.

Notwithstanding many similarities, there are a few basic differences between animal and plant cells in their structure and behaviour. These differences are with respect to cell wall, pigments, vacuoles, centrioles and spindle, turgidity and permeability, cytokinesis, sensitivity to chemicals, extent of experimental polyploidy, energy metabolism, etc. Further, the plant and animal cells differ widely in their responses to various chemicals. Lately, however, there are a few reports of similarity between some plant and animal cells where differences were earlier supposed to be absolute.

The validity of clubbing observations from both plant and animal cells may be known in future. Owing to the differences between these two types of cells, an attempt has been made in this article to discuss the available information on the aberrant responses produced by the meristematic plant cells to chemicals. Studies on animal cells are referred to when necessary and also in cases where comparable information on plant cells is lacking. For a proper understanding of the aberrant cell division, a brief description of normal mitosis and an account of the methodology involved in such studies are thought desirable, since precautions in the latter are often ignored.

CELL DIVISION

Division of one cell into two is a continuous process, but one can distinguish three main phases: (1) Interkinesis or interphase (biochemical preparation for mitosis); (2) Karyokinesis or mitosis (division of nucleus consisting of prophase, metaphase, anaphase and telophase); and (3) Cytokinesis (division of cytoplasm).

Interkinesis or interphase (G1,S and G2 phases). This is a stage of active metabolism of Substances necessary for the division of a cell into two. It consists of (i) Gl (growth 1) phase -presynthetic period, following the preceding division, (ii) S phase' - DNA synthetic period, and (iii) G2 (growth 2) phase - postsynthetic period' preceding karyokinesis. RNA and proteins are synthesized in G1 and G2 periods through 'S'.

In G1 phase, chromosomes are split longitudinally into two strands. During the S period, they separate and each produces a new partner strand. (Therefore, each chromatid of prophase and metaphase chromosomes may be considered to consist of one old and one new strand.) During the G2 phase, physiological changes occur which bring about contraction in the nucleoprotein complex of chromosomes and the initiation of mitosis. A part of DNA synthesis also occurs during this period. The energy needed for these synthetic processes is also generated during interphase. The duration of interphase is variable, although it constitutes the major phase of cell division.

Karyokinesis or mitosis (M phase) - In prophase, chromosomes appear as a network of thin threads. Gradually they become shorter and thicker, revealing their longitudinal doubling into chromatids. Although chromatids are attached only at the centromere region, they neither drift apart nor "become too close. Slowly chromosomes become distinguishable with their characteristic morphology all having a primary constriction (centromere) delimiting the two arms, some having a secondary constriction (nucleolar constriction) also which may be involved in the formation of nucleoli. The presence of matrix around the chromosomes, partially representing material from nuclear sap and nucleol, is controversial. However, they appear to acquire a RNA component during the prophase, which remains, until the telophase, whose origin and fate are not yet known. The forces of condensation of the chromosomes, which bring about reduction in the length and corresponding increase in the diameter of the chromosomes are also not understood with certainty, although at one time it was thought to be due to interactions between negative groups of DNA and polyvalent cations.

Prophase leads to prometaphase with the collapse of nuclear envelope and nucleoli, the former a sudden and the latter a gradual process, followed by the formation of - he spindle. Spindle formation is thought to be due to an interaction between the nuclear sap, nucleolus, centromeres and a clear zone formed just before the disintegration of the nuclear membrane. The scattered chromosomes will then suddenly aggregate at the centre as observed in cinemicrographic studies.

In metaphase, centromeres of the strongly condensed chromosomes lie in the equatorial plate of the spindle. Thus, it represents a stage of equilibrium between the convergent movements of the prometaphase and the divergent movements of the anaphase. The spindle which is responsible for the separation and subsequent segregational movement of sister chromatids is fibrous and is made up of microtubules. These fibres are of three types: (i) extra chromosomal fibres, (ii) centromere fibres, diverging towards the poles, and (iii) continous fibres extending from pole to pole. The nature of polar centres towards which the spindle converges is not known. Divergence of centromere fibres enables them to have more contact with other fibres and to establish crossing points with them, so as to ensure synchrony in the anaphasic movement of chromosomes. Further, the spindle is outlined by mitochondria and is achromatic, for it does not stain with chromosomal stains. It is also considered to be a gel formed by the polymerization of a few types of proteins rich in sulphur. Intermolecular S-S bonds and H bonds play a major role in holding the spindle gel together. Polysaccharides, lipids and even RNA are present in the gel. The ATPase activity is closely associated with the fibrous component of the spindle.

In anaphase, the sister chromatids separate into daughter chromosomes and start moving towards opposite poles with the centromeres followed by the trailing arms. The mechanism of chromosome movement is little known, except that the centromere is essential for the movement. Towards the end of the anaphase, the midspindle expands, resulting in further separation of the two groups of daughter chromosomes.

The events occurring in the telophase are essentially a reversal of those occurring in the prophase. The chromosomes revert to their interphase condition; nudeoli are reformed and the nuclear membrane is reorganized around each group of chromosomes by fragments of endoplasmic reticulum. Nucleolar material is formed on the early telophase chromosomes in the interchromosomal spaces. This prenucleolar material is then collected by the nucleolar organizers (chromosomes with nucleolar constrictions) at the secondary constrictions and thus nucleoli are formed.

The duration of the cvcle of interkinesis and karyokinesis, i.e. the G1, S, G2 and M phases, is variable in different plant tissues. For example, in Vicia faba' it takes 19·3 hr, and in Tradescantia paludosa, 17 hr, most of which is accounted for by the S phase. This mitotic cycle duration is influenced by the nuclear volume and the amount of DNA, except in the case of experimental polyploids.

Cytokinesis (C phase)- As the telophase chromosomal groups are changing to prophase nuclei, the cytoplasm divides into two halves. This is ensured by the formation of cell plate, initiated at the equatorial region, now called phragmoplast equator, and proceeding towards the cell periphery. The cell plate is formed by the fusion of vesicles originated from the golgi apparatus, intercepting the microtubules of the phragmoplast. Fragments of endoplasmic reticulum also are associate with cell plate formation, which probably are responsible for the progressive appearance of neutral lipids, phospholipids, lipoproteins and finally polysaccharides during different stages of cytokinesis. The cell wall slowly develops on either side of the cell plate, which is now left as the future middle lamella.

METHODOLOGY

The extensive literature available on the cytological effects of chemicals shows that the experiments have been carried out under widely different conditions and sometimes without sufficient precautions. Therefore, it is probably proper to stress the importance of methodology and its limitations; though this aspect may be considered elementary, it is often disregarded in the interpretation of results.

Most of the work has been done on the root meristems using seeds of Vicia jaba, Pisum sativum, Allium cepa, etc., by treating them with chemicals directly or indirectly. Seeds offer a uniform tissue in a mitotically dormant state (Go period), and hence are suitable for studies on the initial division cycles. However, one should be cautious about the individual variability among seeds and the age-accumulating aberrations that may later come up and confuse the studies. Keeping a complete record of the history of seed sources is desirable, especially the nature of the soil on which the parent plants were grown. The period of soaking, the seeds should not exceed their saturation level, which varies from species to species. If it exceeds this limit, the substances released from the seeds may exert deleterious effects on them. Excessive soaking may also result in the loss of some important substances from the seeds. Further, handling of the material during germination and throughout the course of the investigations should be under sterile conditions so as to avoid the possibility of pathogenic interference. Before their use in the experiments, the primary and secondary roots should be screened for spontaneous aberrations. The use of meristems is desirable, because they offer: (i) an approximately homogeneous population; (ii) the least individual variability, as it is possible to get several root meristems from the same plant, (iii) a convenient material for employing simple techniques and for administering varying concentrations of chemicals for different durations, and (iv) the least possibility of recording indirect effects due to culture conditions, as may happen in studies with animal cells.

Adequate knowledge of the chemical under study is necessary, such as its solubility, stability in solution, penetration into the tissue and its metabolism. Some chemicals are unstable and get readily hydrolysed, leading to an erroneous understanding of the effects produced. Further, the rate of. Penetration into the tissue could be variable, depending upon molecular sizes and properties and thus may pose difficulties in evaluating the effects, especially in treatments involving short durations. Caution should be taken regarding the technique employed in making cytological preparations, because some dyes like orcein are known to induce the breakage of chromosome. Similarly, several fluids used for metaphase tagging are also known to cause aberrations. Temperature and pH also influence these effects.

Proper designing of the experiment is important for a meaningful study of aberrant mitosis. If effects are observed, direct treatments may be preceded, accompanied or followed by another treatment with 'related' chemicals. This will aid in evaluating the relative ability of these 'related' chemicals to influence the primary events. This is also a way of determining the biochemical antagonisms during cell division, which may yield valuable data regarding the physiology of normal and aberrant cell division.

MITOINDUCTION

Induction of mitosis means initiation of cell division in differentiated cells and/or acceleration of mitotic indices in meristematic cells. Growth promoting substances like indole acetic acid (IAA), naphthalene acetic acid (NAA), 2,4, 5-trichlorophenoxy acetic acid (TCPA) and kinetin are known to induce mitotis in different cells. Therman found a correlation between the period of initiation of the mitotic activity by IAA and the degree of polyploidy in the root tip cells of Allium cepa. Miller et al. and Gutman found that kinetin, in lower concentrations, enhanced the mitotic rate in plant tissue cultures. Sharma and Coworkers induced cell division in the differentiated root cells by ascorbic acid, other vitamins, asparagine, nucleic acids, hormones, inositol and molybdic acid. They reported polyploidy in the differentiated cells and a reduction division of the meristematic cells. Under ascorbic acid treatment increased RNA synthesis was noticed, which perhaps initiated the reduction division.

Earlier, Montalenti, and Brachet also believed that RNA induces cell division. Lately, two new classes of compounds, ureidopurines and nicotinamide derivatives, have been found to have mitoinducing properties. Of these, the nicotinamide derivatives are believed to be the naturally occurring cell division factors, rather than the purine-cytokinins, because of their wide distribution among dicotyledonous plants.

TCPA, IAA and kinetin, which induce differentiated cells to divide, paradoxically cause mitotic depression in meristems by extending the S phase and impairing the G1 cells from entering into the S phase. Van't Hof suggested that certain controlling sites are derepressed in differentiated cells and repressed in meristematic cells by the direct or indirect action of these chemicals.

Several amino acids have been found to increase the mitotic indices in the roots of Ephedra foliata . Colchicine, an alkaloid which is otherwise a potent spindle inhibitor, increased the mitotic indices in several plant meristems by decreasing the interphase duration. This activity, which is similar to that of kinetin, has been repressed by IAA as well as by gibberellic acid, which by themselves are mitoinducers in differentiated cells. It has been suggested that colchicine and IAA affect different stages of cell cycle and that gibberellic acid interferes with colchicine at the transcription level or at the division of centromere.

In the differentiated tissues, where division is stimulated, chromosome reduplication is also generally noticed. According to Mehra, the reduplication occurs naturally during differentiation and the mitogenic chemicals merely render such an endopolyploidy visible. On the other hand,

Bradley and Crane observed that in concentrations lower than that at which cell division is stimulated, the mitoinducing chemicals bring about chromosome reduplication in the interphase nucleus, Later studies have also aroused doubts regarding the validity of the concept that endopolyploidy accompanies or precedes cell differentiation.

The exact mode of mitoinduction is little understood. One of the earlier explanations was that the DNA of differentiated cells is deficient in sugar moiety and by supplementing such cells with DNA, division was stimulated. Sharma and Chatterjee suggested that the mitoinducing chemicals may also alter the nucleic acid-protein balance or accelerate certain cellular activities which haveceased to function during differentiation. Recently, Bekhor et al. have separated protein and DNA from isolated chromatin, so as to find out the conditions' under which their exact or equivalent rebinding can be brought about. Such studies on the chromatin of meristematic and differentiated cells may reveal the nucleic acid and protein changes during differentiation. Thus, an understanding of the process involved in the gradual cessation of cell division may help in explaining the induction and acceleration of mitosis.

An interesting aspect of mitoinduction is the problem of mitotic trigger, which can switch on and off the division process. It is known that nuclear division is always synchronous in multi-nucleate cells, meaning thereby that a trigger exists and is located in the cytoplasm. Even the partial synchrony commonly found in meiotic systems is known to be due to intercellular connections and hence is suggestive of the location of the trigger in the cytoplasm. An induction of asynchronous division in multinucleate cells and synchronous division in meristematic cells may be significant vis-a-vis the cessation or induction of division in differentiated cells. Such mitotic synchrony has been obtained to a small extent in the meristems, of Pisum sativum with 2,4-dinitrophenol (DNP) and 2,4-dichlorophenol (DCP) and to a greater extent in Vicia faha with 5-aminouracil (AU). AU initially inhibits RNA synthesis and later retards DNA synthesis. When the chemical is removed, a sudden flush of synchronous mitotic activity appears.

MITOINHIBITION AND MITODEPRESSION

Mitosis can be reduced in frequency or even completely inhibited. A large number of chemicals are known to have such depressing or inhibitory properties and some of them may even affect other stages of mitosis. Protoanemonin, antibiotics like streptomycin and actidione, nitrogen mustard, benzimidazole and diaminopurine, the sodium salt of DNA, coumarin, acridines, herbicides, such as pentachlorophenol (PCP), trichloracetic acid (TCA) and aminotriazol, are some of the wellknown prophase or preprophase inhibitors.

The enumeration of cellular processes affected by the chemicals resulting in mitotic inhibition is bound to be arbitrary due to inadequate knowledge of the biochemistry of cell division. Mitosis initially requires a highly well-balanced metabolism and naturally this equilibrium can be affected in a great number of ways, such as division of the centromere, which may be chemically different from the rest of the chromosome; uncoiling of the nucleoprotein strand during chromatid separation; organization of spindle; cell cycle; release of energy; cytoplasmic clotting mechanisms; and a possible synthesis of a trigger substance.

Some information is available concerning the agents acting at some preprophase stages. Several sulphides and sulphocyanides block the cell from entering into the S phase. A few metallic ions (Sc^{3+},A^{13+}) and glutaminic acid also affect the S phase. Maleic acid hydrazide (MAH) inhibits RNA synthesis by interacting with the enzymes responsible for it. Protein synthesis, in general, occurring during the cell cycle, is shown to be inhibited by cycloheximidelo. The RNA synthesis at the G2 phase, and early prophase are inhibited by mitomycin C, Ethidium bromide, 3'-AdR and ribonuclease, thus blocking cells from entering the prophase and the metaphase. Further, it was reported that RNA causes inhibition of cell division. A likely explanation for such an inhibition is that foreign RNA, which is chemically related, but not identical with the cellular RNA, competes or interferes with its activities. In other words, the foreign RNA might play the role of chemical analogue of the cellular RNA.

More often there may be a decrease of mitotic indices to different degrees than a omplete inhibition. Such mitodepression has been caused by lauryl a1cohol, different isomers of hexachlorocyclohexane, phenols, propanil and its degradatbn products. Rifamycin depresses mitotic indices by prolonging the RNA synthesis. Similarly, 6-methyl coumarine (MC) delays the S period slightly and the G2 period considerably. In general, most of the antichromosomal and antispindle agents are mitodepressive.

The processes of mitodepression and mitoinhibition are related in that their differences are of degree. The inhibitory substances may act only at the interphase or at both the interphase and the prophase. Consequently, several variations of depression and inhibition are noticed, viz. (i) the number of cells entering the prophase decreases, while the post-prophase mitoses proceed regularly as seen in the case of roots treated with maleic hydrazide, (ii) both the onset of prophase as well as the post-prophase mitoses are affected, as in the case of tissues treated with isoniazid compounds. When the chemicals act at the interphase, one can specify where the mitoinhibition occurs: before, during or after DNA synthesis (i.e. G1 S or G_2 phases respectively). But the mitodepressive effect need not always be due to a specific impact on any of the above phases. When the inhibiting chemicals are removed, the recovery may follow immediately, as in the case of tissues treated with antipyrenes, or slowly, as with antifolates. Further, it is important to find out whether the dividing cells are those recovered from inhibition or those not inhibited at all.

The differences in the actions of various chemicals on mitosis enumerated above reflect variations in the biochemical mechanisms of mitodepression and mitoinbibition. Recently, a simultaneous impairment of golgi vesicles accompanying mitoinhibition has been revealed through electron microscopy. Therefore, it is possible that the proper functioning of dictyosomes is an essential prerequisite for the onset of mitosis and any effect on this may indirectly result in mitodepression or mitoinhibition.

CYTOINHIBITION

Cytokinesis. is distorted or prevented by some chemicals which interfere with the normal development and functioning of the phragmoplast. Simonet and Guinochet first found that the halogen derivatives of benzene and toluene have this kind of effect on the cells of Linum. A similar response was elicited by adenine and methylated purines like theobromine and theophylline from the root meristems of Allium cepa. Wada reported that 2-hydrazinotropone and l-bromo-2- hydrazinotropone in lower concentrations do not impede mitosis in Tradescantia, but dissolve the phragmoplast, thereby causing binucleate cells. Other substances like aminopyrine and some Vinca alkaloids also cause cytoinhibition. Some hormonic and nonhormonic herbicides have been found to cause complete inhibition of cytokinesis and partial or multiple cell plate formation.

Digitonin selectively interferes with the polymerization of phragmoplast microtubules, resulting in the inhibition of cytokinesis, without interfering with spindle microtubules. Rozmus found that the sodium salt of DNA prolongs the mitotic cycle without causing mitodepression. But anaphases and telophases were found to be irregular and the cytokinesis was delayed. Thus, there seems to exist some interrelationship between the interphase activity, spindle and cell plate formation.

Gonzalez using lindane and caffeine inhibited cytokinesis in two successive cell divisions and obtained bi-, tri- and tetranucleate cells in the roots of Allium cepa, the processes being termed as bimitosis and polymitosis. The nuclei divide synchronously and may even get fused, resulting in polyploidy. The duration of the cell cycle in such multinucleate cells has been found to be increased by 15%, most of which is solely due to the M phase. There is no change in the durations of the Gl, s and G2 phases. Based on these studies, Gimenez-Martin et al believe that the mechanism regulating the initiation of mitosis may be located in the cytoplasm.

Cytoinhibition was initially considered to be due to the destruction of the cell plate. But it appears that cell plate precursors produced by golgi vesicles are destroyed. In fact, any of the processes involved in cell plate formation, viz. production and aggregation of golgi vesicles, their arrangement in the equatorial region and coelesce to form the cell- plate, could be affected. Even cytoplasmic currents responsible for vesicular movements may get modified. On the other hand, Belousova et al suggested that the oxidative phosphorylation in the mitochondria may be inhibited, thus affecting the ATP pool. Sometimes the cell plate instead of being inhibited is merely shifted away from the equatorial region. This could be either due to the dislocation of golgi vesicles from their normal position or disruption of the cellular equidistance of the telophase groups caused by interference with the expansion of the midspindle. Concerning the consequences, inhibition of cell plate formation may' affect the nucleocytoplasmic balance and render the cell incapable of further divisions. Such an eventuality may be useful in the cessation of tumorous growths. Induced binucleation can also be a valuable method for studying the cell cycles.

ANTISPINDLE EFFECTS

A large number of organic and inorganic chemicals disrupt the spindle to different degrees and even inhibit it, causing polyploidy. Ostergren enumerated 80 such substances and the list continues to grow. A few important ones are: mineral and organic derivatives, organoaliphatic compounds like chloral hydrate, ether, chloroform, various alkaloids-coumarins, acridines, antibiotics like griseofulvin, organic mercurials purines like caffeine and theophylline and a wide variety of other compounds. Mild antispindle effects have been recorded for phenols, mimosine and several herbicides. Some of the herbicides have been recommended even for routine chromosome clarification studies.

Of all the antispindle chemicals, the most well known is the alkaloid colchicine, which occurs in the corms of Colchicum atumnale and C. Luteum and a few other species of Liliaceae. Colchicine disrupts only the spindle of the dividing cells of any group of plants and has almost no effect on the other parts of the mitotic cycle. Consequently, the chromosomes fail to move poleward and remain confined to the equatorial region. This is termed C-mitosis, which ultimately leads to polyploidy.

Interestingly, the species of Colchicum are immune to colchicine, although some reports to the contrary are available. The prothalli of some ferns treated with colchicine produced tetraploid sperms and some of them are even monstrous. Since the first application of this chemical by Blakeslee and Avery, a new vista in biological research, from both utilitarian and fundamental standpoints, has opened up.

Podophyllotoxin, an alkaloid obtained from Podophyllum peltatum and P. emodi, has been found to be effective in much higher dilutions than colchicine. Yet colchicine is universally employed for experimental induction of polyploidy owing to its high water solubility and low toxicity at effective concentrations, enabling speedy recovery of the affected cells. The cells of angiosperms, gymnosperms and lower plant groups are similarly affected by this compound. Other important antispindle agents deserving special mention are griseofulvin, vinblastine and vincristine. Griseofulvin, which is extracted from the broth cultures of Pencillium griseojulvum, enhances the action of colchkine when used along with it. Vinblastine and incristine, alkaloids from Vinca rosea, are well known for their antineoplastic properties, mainly because they depress and inhibit mitosis completely after poisoning spindle formation. Their other related alkaloids such as vindoline and lochnerinine are also potential antispindle agents.

Following the administration of an effective concentration of colchicine, C-mitosis ensues and its constituting phases are called C-metaphase, C-anaphase and C- telophase. C-metaphase is much longer than the normal one, where the chromatids are shorter and do not separate. In C-anaphase, the centromere junction breaks, but the daughter chromosomes lie in characteristic pairs. In C-telophase, the 4n chromosomes are drawn together and an irregular nucleus is formed. In plants, C- mitosis is iterative, i.e. the cells can undergo successive C-mitoses. In C-metaphase, the chromosomes may lie in a fashion appearing either as a ball or as a star or scattered irregularly or distributed in groups of two. If the distribution of the chromosomes is even, it is designated as 'reductional mitosis' and has been obtained with antibiotics and the sodium salt of RNA. Sometimes the chromosomal groups may be more than two with an uneven number of chromosomes.

Star or ball appearance may be due to the arraying of unseparated centromeres at the equatorial region, as has been observed only in some animal tissues. Such star metaphases degenerate by pyknosis. Further, polyploidy is rare and C-mitosis is not iterative to any extent in animal cells. These major differences between the animal and plant cells are supposed to be related to a blocking of RNA synthesis during mitosis and a more rapid disappearance of a 'needed' messenger RNA in animal cells. A slow but definite effect of colchicine on centromeres has been observed by Deysson.

Of the many colchicine derivatives studied so far on Allium cepa, only aminocolchimidazole has been found to have no C-mitotic effect. Deacetyl aminocolchicine is the most active and even causes mitoinhibition under certain conditions. Based on his extensive work, Deysson concluded that substitutions on carbon atoms 14 and 15 determine the antimitotic properties and probably it is at these sites that the colchicine molecule links to a cell component.

The minimal effective concentrations of some antispindle substances are given in a Table. It is found that the animal cells are more sensitive to antispindle agents than the plant cells, and hence the latter require higher concentrations to produce the desired effect

Although a vast amount of literature has accumulated on colchicine effects, the exact biochemical mechanism of colchicine action is still not well understood. However, many views have been put forward from time to time. Before discussing them, some important recent observations on colchicine action may be cited.

Bajer and Mole-Bajer found that the clear zone which is the first sign of spindle formation begins to grow in C-mitotic cells only after the disappearance of nuclear membrane, unlike in normal mitosis. During their studies on mitosis in the living cells employing cinemicrography, they found that chromosomes have a role in the formation of this clear zone. Thus, colchicine action can be twofold: (1) directly on spindle structure, and (2) indirectly on the chromosomes, preventing subsequent development of the spindle.

Another important observation is the disappearance of microtubules in' the spindle during C-mitosis, like those of the cell wall. The wall micro-tubules are considered to ensure the Isodiametric cell growth, because they disappear from the walls of colchicine-induced root tumour cells. However, the role of microtubules is still imperfectly understood, and their responses to spindle inhibitors may be variable. Sato found that colchicine reduces the microtubules to granules, while mercaptophenol, another spindle inhibitor, merely renders them sticky.

The above observations indicate that colchicine acts at definite points in the mitotic cycle resulting in the inhibition of spindle formation. This action disappears when colchicine is removed. It has no impact on an already organized spindle and has some effect on the chromosomes which may indirectly be connected either with spindle formation or function. Further, colchicine is modified as soon as it penetrates the cell. Besides these implications, one must consider the differences in the C-mitosis of the plant and animal cells for a proper evaluation of various theories advanced to explain the biochemistry of C-mitosis. These theories relate to colchicine action on proteins, RNA, ATP, m-inositol, enzymes, the oxidation reduction processes and the electron transport chain, which are briefly described below.

Action on proteins - C-mitosis has been considered to be due to the superprecipitating action of colchicine.It is known that colchicine interferes with sol-gel transformations in both mitotic and nonmitotic cells,and inhibits the polymerization of microtubules, which are prominent structures of spindle and ultrastructural markers of gelated systems. Just like griseofulvin and vinblastine colchicine interferes in cells having an organization of labile fibrillar systems (protoplasmic gels). In some cells it is also known to inhibit the release of histamine whose role in mitosis is still not c1ear. Studies using tritiated colchicines have shown that colchicine appears in a soluble protein fraction which is not a major component of the spindle, but may be a subunit of the microtubules. Wilson's studies revealed that odophyllotoxin, mentioned earlier, inhibits the binding of colchicine to protein in contrast to Vinca alkaloids and that the above inhibitory effect is not brought about by a simple competitive mechanism. These findings raise two important questions: Whether all these C-mitotic chemicals have a common mechanism of action or each alone, particularly colchicine, has several mechanisms of action in cells. However, the cell proteins to which colchicine binds and their possible relationship to microtubules are of considerable importance. Changes in he electrical charges of the micro-tubule protein or a rearrangement in the molecules with changes in their secondary or tertiary structures cannot be ruled out as contributing to the mechanism of C-mitotic action.

Action on RNA - Based on UV spectral studies, Vles and Tatar-Goldberg considered that colchicine acts on RNA. Some evidence in favour of this opinion has been lent by the recent work of Kleinfeld and Sisken. They found a block in RNA synthesis and disappearance of a much needed messenger RNA, which is generally more rapid in the animal cells than in the plant cells.

Action on ATP — According to Brachet, ATP provides energy to the spindle for its activity in the movement of chromosomes. Hence, the inhibition of the spindle is supposed to have been brought about by the reaction of colchicine with ATP. An antagonism between ATP and colchicine has also been claimed.

Action on m-inositol- m-Inositol plays an important role in the mitotic events. Some antagonism has been found between it and colchicine. Hence, it was thought that colchicine may act on m-inositol for bringing about spindle inhibition.

Action on oxidation-reduction processes - As early as 1938, Gal found that colchicine inhibits several dehydrogenases. Galzigna and Cristofolini and Galzigna found that colchicine stimulates the oxidation of glutathione, one of the SH-rich spindle constituents. They regarded spindle gelation to be dependent on the S-H-SS cycle. Similar effects on other oxidation-reduction processes and anatagonisms between colchicine and ascorbic acid dimercaptopropanol and mercaptocthylamine are also known. All these point out that the oxidation- eduction processes which are vital for normal mitosis are affected by colchicine.

Multiple action - Chakraborty and Biswas reported an increase in RNA, DNA. and protein contents and a decrease in ATP and nucleotide pool in the C-mitotic cells of Allium cepa roots. They also observed a decrease in the free sulphydryl content and a corresponding increase in disulphide groups. These findings reflect a major change in the configuration of spindle proteins.

Action on enzymes - It was reported that colchicine brings about the inhibition of deoxyribonuclease activity, acid and alkaline phosphatase activities, proteogenesis and also stimulation of glucide metabolism, indicating that several enzymes and a few miscellaneous activities are affected by colchicine.

Effect on electron transport chain - Deysson and Delage brought about protection from colchicine effect in Allittm cepa by applying yeast extract prior to colchicine treatment. They traced this protective effect to riboflavin and nicotinamide,which are present in several coenzymes operating in the electron transport chain. Based on these experiments it was concluded that spindle inhibition might have been brought about by the blocking of any of the links in the electron transport chain.

But vitamin protection against spindle poisoning is not universal. Yeast extract is as protective against vinblastine as against colchicine, less protective with griseofulvin and phenylurethane and completely ineffective with chloral hydrate and antipyrene. Thus, the mechanisms of spindle inhibition by different chemicals can be different, although the end result may be the same, and hence any single explanation may not be universally applicable.

A few other spindle derangements are caused by antispindle agents Barthelmess reported the induction of multipolar spindles by 'athnol' in Allium cepa. This resulted in irregular chromosome segregation and the formation of cells with nuclei of different sizes. Irregular chromosome dispersal within the cell has been obtained by employing calcium antagonists. Lauryl alcohol, organic mercury halogenides and several terpenes also induce bi- or multipolar spindles. The action of terpenes is related to their structures. Fernandez-Gomez et al found that the delta isomer of hexachlorocyclohexane (HCCH) can produce multipolarity and weak C-mitosis, while its other isomers can only be mitodepressivello. Thus, the structures of compounds also may be the determining factors in inducing certain types of aberrations.

Ultrastructural studies of multipolar spindles induced by chloral hydrate revealed that the microtubules of the continuous fibres are destroyed, while those of the centromere fibres become convergent. Divergence of the centromere fibres is essential for synchronous bipolar movement. Lack of such divergence naturally disturbs both bipolarity and synchrony. Recently, based on his work on Agropyron, Tai proposed a theory according to which the spindle organizer (polar centre) of plants is a unit cell organelle and is similar to the centrioles of animal cells. If so, any chemical which can affect the polar centre may cause multipolar spindles and efficiently reduce the ploidy level of cells. However absence of information about a specific effect on centrioles and reduced ploidy through spindle multipolarity casts doubt on the validity of Tai's theory.

Gradual topographical shift of the spindle from its usual longitudinal axis to the transverse axis has been caused bv some amino acids and alkaloids. The significance of this phenomenon which is called topokinesis is not known, except that it may be of some histological consequence. The exact mechanism of shift in the spindle orientation is obscure. Probably, the situation of polar centres towards which the spindle fibres converge might have been affected. This naturally influences the orientation of the telophase chromosomal groups and subsequently the plane of cell plate formation.

ANTICHROMOSOMAL EFFECTS

Chromosomes break up into fragments on treatment with some chemicals, just as they do when they are subjected to radiations. Such chemicals are said to be 'radiomimetic'. The term is controversial, because the chromosomal breaks caused by chemicals are generally independent of oxygen delayed in appearance and localized in occurrence unlike those produced by radiations. The term, however, has never been used strictly, except in a limited sense, because of wide variations in the type and mode of production of breaks by different chemicals.

The action of such chemicals may be primary when the free chromosomes of mitosis are affected (physiological, non-delayed, G2 or M phase aberrations), or secondary when the chromosomes at the interphase are affected and the effects become visible only on recovery (structural, delayed, G1 or

S phase aberrations). However, the production of the effects depends upon the ability of the chemical to penetrate the nuclear membrane. Based on this, for example, Kihlman divided purines into two groups - the lipid soluble and the water soluble. The lipid-soluble ones penetrate the npclear membrane and bring about secondary aberrations in the chromosomes; the water-soluble ones, on the other hand, cause breakage during mitokinesis. For example, mustards which are lipid soluble have secondary effects acting at the interkinesis, while acridines which are water soluble have both primary and secondary effects. Breaks may occur simultaneously in the two cbromatids of a chromosome or only in one chromatid. Further, there is some evidence indicating the occurrence of subchromatid breakage restricted to one or more strands of the chromatid.

Although chemical effects are nonrandom, they may not always be specific. Further, when the effective concentration is low, nonrandomness tends to disappear at longer durations of treatment. Nitrogen mustard and diepoxypropyl ether (DEPE) have produced the largest number of aberrations in the heterochromatic regions of the long arms of the short chromosomes in Vicia jaba. Maleic hydrazide (MH) has a preferential action on heterochromatin on either side of the centromere of the long satellite bearing (SAT) chromosomes; 8-ethoxycaffeine (EOC) also acts on the same chromosome, but in the nucleolar region. Similar localized breaks have been produced by mitomycin C and several sulphonates.

Summarizing the observations of his colleagues, Oehlkers states that the action of ethyl urethane is chiefly on the centromeric and nucleolar regions of the SAT-chromosomes in *Vicia faba.* According to McLeish, the differential specificity of DEPE and MH is due to their action on different systems of chromosome synthesis. The systems may be blocked by these chemicals at different stages. According to Marquardt[2], there may be, in addtion, some sensitive regions in the polypeptide chain of the chromosome arms where breaks can occur more frequently and easily. Revell suggested that possibly some of the labile sites along the chromosome length, which normally break at the meiotic crossing over, are more susceptible to breakage during mitosis. Unfortunately, hetero-chromatic pairing is not universal for all biological systems. However, heterochromatin-localized breaks may indicate its general susceptibility and differences in base composition along the chromosome length. Wuu and Granthave recently produced a variety of breaks with several biocides, the most important being those localized at the satellite whose significance is obscure. Recently, Sing *et* al obtained nonrandom chromosome breaks, mainly localized in the centromeric regions in barley with lysergic acid diethylamide (LSD). However, Sturelid and Kihlman could not find any such localized breaks in *Vicia faba.*

As a result of these types of breaks, there may occur deletions of parts of chromosomes, formation of acentric fragments, misdivisions, sister reunion of chromatids, chromatid reunion, translocations, formation of pseudochiasmata, dicentric chromosomes and anaphasic chromatid bridges. If the changes are of a drastic nature, the genome becomes unbalanced and the affected cells are rendered incapable of further division and may eventually perish. In fact, the chemotherapeutical treatment of cancer is based on this principle, and hence there is constant search for newer chemicals with the hope of finding one which destroys the cancer tissue without harming the normal tissue of the organism. In addition, the translocations resulting from chromatid reunions, if viable, cause heritable phenotypic differences which are important from evolutionary standpoint. Further, the differential nature of the chromosome segments even between the primary and secondary constrictions and hetero- and euchromatin is revealed by the variable activity of the radiomimetic chemicals on the chromosomes of the same organism and among different organisms.

Urethane is the first chemical found to mimic radiations in causing chromosome breakage in the root tips of *Vicia faba.* Since then extensive work has been done on the radiomimetic properties of chemicals. From such data, however, deductions about the mode of action of these chemicals are extremely difficult. Prominent of these chemicals are: (i) Alkaloids of various types (ii) hydrocarbons and azocompounds, (iii) naphthalene derivatives, (iv) acridine derivatives, (v) phenanthrene derivatives, (vi) halogen derivatives, (vii) coumarin derivatives, (viii) salicylic series, (ix) Mercurials (x) nitroso compounds, (xi) heterocyclic bases, (xii) drugs and bacterial products, (xiii) mustards, (xiv) phenolS, (xv) hormones and other growth substances (xvi) natural products like vegetable oils, pigments, oils and essences, bark and seed extracts and (xvii) a great variety of other chemicals.

Among the metallic compounds, mercurials like phenyl mercuric hydroxide and phenyl mercuric nitrate and heavy metal salts like aluminium chloride and cadmium nitrate in extremely low concentrations are shown to be radio-mimetic. Von Rosen brought about chromosome breakage in *Pisum sativum* employing strong complex-forming metal ions like Cu^{2+}, Ag^{+}, Au^{3+}, Zn^{2+}, Cd^{2+} and Hg^{2+}, which are normally required by plants only in trace amounts. Based on these results, he put forward an interesting theory that varying concentrations of such metal ions in soil may have played a vital role in evolution by inducing natural mutations in plants and indirectly in animals feeding on them. In fact, he has arranged all the compounds studied by him in series based on their relative radiomimetic efficiency. The series include, besides the metal ions mentioned above, halogen series, alkalies, acids, salts, organic compounds and other substances.

Cupferon, a chelating nitrosamine, produced delayed isolocus breaks of chromatid type in *Vicia faba*, which could be modified by pH and temperature variation. Its action is supposed to be due to the inhibition of respiration in the ce11. Sodium nitrate, diphenyl nitrosamine (DPNA), nitric oxide and N-methyl phenyl nitrosamine (MPNA) also produced similar effects. But the breaks caused by MPNA are localized near' heterochromatin and are independent of pH and temperature.

Merz obtained abnormal metaphases, isolocus breaks and exchanges by mitomycin C and azaserine. Streptonigrin, another antibiotic, has been reported to produce pseudo-chiasmata, chromatid fragments, chromatid bridges and chromosome exchanges in *Vicia faba.* Most of these effects are localized in the nucleolar and centromeric regions. Similar localized effects of nondelayed type were observed by

Davidson with azaserine and the action is attributed to the effect on purine and protein metabolism. An interesting idea has been put forward by de Carvalho that the radiomimetic properties of the bacterial products render the soil mutagenic. Thus, soil becomes a potential source of spontaneous chromosomal aberrations which are vital for the evolution of species.

Breaks in chromosomes often accompany the induction of division by hormones· These breaks may be due to a tension during chromosomal separation or due to nonsynchronous chromosomal replication and separation. According to Sharma and Sharma, hormones cause karyotype clarification and fragmentation, below and above the threshold concentration required for growth. Thus, even a metabolite, in excess amount within a cell, can cause aberrations.

Mustards, nitrogen as well as sulphur types cause significant breaks, just like their other related alkylating agents, viz. ethylenimines, epoxides, sulphonates, etc. All these contribute a positively charged alkyl radical, which may act at several points in cell metabolism and even denature DNA by crosslinking guanine in the two strands of DNA. Some other substances like mitomycin C and azaserine, mentioned earlier, also act in this manner. The pseudochiasmata, commonly found in the above cases, preceding the breaks, may, therefore, be formed as a result of the delayed division of certain chromosome segments.

Weak radiomimetic effects are exhibited by several heterocyclic bases, adenine, guanine and uracil, in *Allium cepa, Pisum sativum* and *Vicia faba.* These bases inhibit DNA precursors in the cells, which have not completed DNA synthesis. No aberrations involving rejoining of breaks occurred because DNA synthesis is required for the repair. Further, the toxic effects of purine derivatives increase if methyl and other alkyl groups are introduced. Thus, 8-ethoxycaffeine produced isolocus breaks and chromatid exchanges of non-delayed type, similar to radiation effects. Other purine analogues like 6-mercaptopurine and azaguanine inhibit the formation of thymidylic acid from deoxyuridylic acid.

Similarly, an antipyrimidine, 5-fluorodeoxyuridine (FUdR), produced chromosomal, chromatid and isochromatid gaps of the non-delayed type, while many of its related compounds have no effect on the plant cells. FUdR and bromodeoxyuridine (BUdR), which are structurally similar, act differently. The former inhibits the DNA synthesis and the latter replaces thymidine in newly formed DNA. Therefore, structural similarity may not necessarily imply similarity in action. Thus, the antimetabolites may inhibit the synthesis of purine and pyrimidine bases or act as substitutes for them.

Unlike in the case of X-rays where the dosage and breaks are directly proportional, in the case of chemicals there is a definite optimum concentration below and above which the number of breaks falls, although other effects may become visible. Further, mutual antagonism is exhibited by some chemicals. For example, urethane and aluminium chloride, which are individually effective in producing chromosome breaks, nullify the effect of each other when used together.

The numerous chemicals that produce antichromosomal effects show a wide variation in their effective concentrations.

Some of them are very active in low molar concentrations. Most of the agents, in general, have mitodepressive and mitoinhibitive properties in higher concentrations, while a few of them have antispindle properties also. The ratio between the lethal concentration and the mitodepressive concentration is very important. When this ratio is low, it is not possible to differentiate between the antimitotic activity and cytotoxicity, because the antimitotic effect could be a manifestation of the cytotoxicity of the chemical.

The animal cells appear to be more sensitive to antichromosomal agents than the plant cells, except in respect of the full mitoinhibitory effect of triethylene thiophosphoramide (TEPA). Further, the plant cells respond quickly, while the animal cells in culture get blocked in the metaphase for longer periods and show spontaneous aberrations, which may hence confuse the studies.

Generally, chromosome breakage is a reflection of the break in the nucleic acid molecule and the causative factors may be different, although the end result is the same. These factors include impairment of the sulphydryl groups of proteins, the hydrogen bonds of nucleic acids and the oxidation-reduction system within the cell. Further, there could be interference with RNA/DNA ratio in the chromosome to such an extent that breakage occurs in segments, which are involved in RNA metabolism. This happens especially in the case of heterochromatin-localized breaks. Some other possible means by which chromosomal breaks might originate include the removal of DNA-bound metals and denaturation or defective synthesis of nucleoproteins. In fact, the biochemical mechanism of induction of breakage is known only for a very few compounds.

The interactions in the mitotic cells at the chromosomal and physico-chemical levels, during and after the exposure to antichromosomal agents, are also given.

The events shown indicate the possible ways by which chemicals can act on the chromosomes. It would appear that the mustards and diepoxides, being highly reactive, undergo chemical changes non-enzymatically before they enter into the biochemical phase of activity. MH and EOC, on the other hand, being derivatives of nucleic acid bases, act as antimetabolites and immediately enter the chain of events through the biochemical pathways. All the radiomimetic chemicals eventually get involved in some part of the oxidative machinery of the cell. It appears, therefore, that interference with any part of the oxidative metabolism of the cell is vital for both induction of and recovery from aberrations.

At the chromosomal level, the two major categories of events are breakage and rejoining, and each of these is further subdivisible. The breakage, whether of chromosomal, chromatid or subchromatid type, may be potential or actual. The rejoining system can be divided into restitution and recombination phases, which are competitive processes for the disposal of chromosome fragments. A great majority of the breaks undergo restitution, which may involve repair of potential breaks or the rejoining of actual breaks or both. Recombinations will always result in the reshuffling of the genotype.

In the case of radiation-induced breakage, rejoining could be effectively separated from breakage. For example, in *Tradescantia,* chromatid breaks remain open for 5 min and chromosome breaks for nearly 1hr for recombination. But in chemical breakage these cannot be separated.

It is evident that fragmentation occurs following the action of a variety of chemicals differing widely in their properties. Based on the end result alone one cannot deduce the mode of action of these chemicals and assume similarity of mechanism of their actions. To find out the role of metabolism in breakage, experiments with protective and/or recovery chemicals, if any, are needed, as was done in the case of X-rays. Recovery from injury caused by any particular chemical may indicate the affected metabolic step and the deficiency caused by such a chemical. Very meagre work on these lines has been done so far. In this connection mention may be made of the reversal of azaserine effects by amino acids and the influence of cystine on the effects produced by nitrosoguanidine. Sometimes the pre or post-treatments of the treated cells by other substances may enhance the effects of the primary agent. Thus, post-treatments with cycloheximide and chloramphenicol increased subchrornatid exchanges among the breaks induced by 6-methyl coumarin.

CONCLUSIONS

Broadly speaking, two major consequences emerge from the various studies on antimitotic effects besides throwing some light on the linking up structure, function and visible effect within the cell. The first is a disturbed metabolism whose subsequent impact on growth can be to any extent and degree. The second, which is more important from a cytogenetic point of view, is the shuffling of the heredity material. Such a reshuffle may result either in polyploidy whose role and importance in increased vigour of plants is well known, or in mutations, the effects of which are more harmful than beneficial.

Systematic screening of newly synthesized compounds for antimitotic properties may become important in view of the increasing human consumption of an evergrowing variety of accessory foodstuffs and drugs. The study of antimitotic effects may also aid in understanding the permeability variations among the living systems to different chemicals. When adequate information is available on the biochemistry of normal and aberrant mitosis, data on the concerned metabolic events may be of immense practical value. Further, as Deysson remarked, "every impairment of the metabolism will be shown up as a visible morphological effect on mitotic proliferation, rate or characteristics; the study of antimitotic effects may constitute a technique suitable to various problems of general physiology".

Future work may centre round the antimitotic effects of chemically undefined natural products like seed extracts. When such extracts are taken at different periods, valuable data concerning ageing as well as antimitotics and cytotoxicity may be obtained. The substances accumulated in seeds during ageing, besides influencing their germination and further growth, may have some pharmacological value. The work on aberrant mitosis may also reveal the nature of immunity present in some plants to the substances produced by them, which are otherwise antimitotic in other plants.

Further, the selective activity of several substances. in different plant and animal cells raises the question of drug resistance in dividing cells. Another interesting line of work deals with the recovery and protection of cells from antimitotic effects making' use of antimetabolites and chemical antagonists. However, the information available on this aspect is scanty. The study of the cytogenetic effects of biocides and their degradation products is also of considerable importance. Due to their gradual, accumulation in soil in different forms and concentrations, the crops and other important plants may be genetically affected in due course. Therefore, screening of biocides, their active constituents and degradation products for antimitotic properties is of considerable practical utility.

4

ENVIRONMENTAL MUTAGENESIS-I

THE BEGINNINGS CHEMICAL MUTAGENS AND THE ENVIRONMENTAL MUTAGEN SOCIETY

By Samuel S. Epstein, M.D

After obtaining a Doctorate degree from BITS, Pilani, I got an opportunity to work in the IISc, Bangalore towards the end of 1969. I was groping in darkness not knowing what to do. And I am made independent too soon! The best way is to spend in library, talk to as many as possible relevant to Biology and attend Lectures. Thus I chanced upon crashing into a personality – R. Nagaraja Rao, who did Doctorate from IARI, New Delhi and joined here as a Research Associate after a stint in USA as a Post-Doc. He suddenly shouted **"Why are you so diffident - you worked on Vinca Alkaloids – the whole field is open – People like us formed a new Society in USA – join there and follow".** *My shadows are dispelled. He gave some literature. Others I obtained through mail. Some of them are the inaugural issues of EMSUSA News Letters. In the very first issue is the below reproduced article by Epstein, the very first Secretary of Environmental Mutagen Society of USA.*

I was lucky to have met him personally at Penang in Malaysia in 1986 during an International Conference on Human Future. Infront of him I pronounced my theory of Gene-Epidemics from Pesticides. After going to US he sent me his book on the Mutagenecity of Pesticides, a publication of early '70s.

His article of 1969 is reproduced here for the reference of the interested people.

First Article on Environmental Mutagenesis by the First Secretary of EMS, USA, 1969
Reproduced with the approval of the Publisher, reprinted from The Medical Tribute.

For many years, ionizing radiation from external sources and radiation from absorbed radioactive substances have been known to damage genetic material. The need for appropriate detection and control measures has consequently been widely accepted and the limitation of human radiation exposure to lowest possible levels is now a well-established public health principle. The additional possibility that chemicals may produce adverse genetic effects in man has not previously received general recognition, although geneticists and molecular biologists have for three decades utilized various chemicals for characterizing the genetics of specific systems and have shown that these chemicals can induce genetic damage similar to radiation.

Concern is now growing that chemical mutagens may pose a potential public health hazard. Such concern has been emphasized in a recent NIH genetic study section report (Crow, 1968), which states: "There is reason to fear that some chemicals may constitute as important a risk as radiation, possibly a more serious one. Although knowledge of chemical mutagenesis in man is much less certain than that of radiation, a number of chemicals --some with widespread use — are known to induce genetic damage in some organisms. To consider only radiation hazards is to ignore what may be the submerged part of the iceberg." The report further states: "Recent investigations have revealed chemical compounds that are highly mutagenic experimental organisms, in concentrations that are not toxic and that nave no overt effect on fertility..."

The main categories of chemicals in which mutagenicity may be anticipated include alkylating and radical-producing agents, nucleotide analogues, and compounds that interfere with nucleic acid synthesis. Ethyleneimines are examples of highly mutagenic alkylating agents. [These are used for a wide variety of non-industrial and industrial purposes, including therapy of neoplastic and autoimmune diseases, chemo sterilant insecticides, pigment dyeing and printing, fireproofing and crease proofing of fabrics and textiles, ingredients in jet and rocket fuels, polymer stabilizers and accelerators, intermediates in a wide variety of industrial syntheses, and as cross-linking agents in starches and shampoo. The use of these compounds has even been recently recommended for rodent and plant control!

Mutations can produce a wide range of deleterious effects. Dominant mutations express themselves immediately by production of early fetal deaths, and abnormalities in the F1 generation such as Achondroplasia, Polydactyly, and Retinoblastoma. Recessive mutations such as albinism, Fanconi's anaemia, amaurotic idiocy and phenylketonuria, may not express themselves for many generations until both parents with identical genetic lesions happen to bear homozygous offspring. It is likely that the major effects of increased mutation rates would be spread insidiously over many generations, and would include ill-defined abnormalities, such as premature aging, alterations in sex ratios, and increased susceptibility to various diseases, including leukaemia and cancer.

Fortunately, a wide range of screening and other more definitive tests are now available for characterizing potential chemical mutagens. Screening systems include physicochemical and transforming tests with isolated DNA, classical microbiological tests, such as those based on the adenine-3 locus of Neurospora and the histidine operon of Salmonella typhimurium, the mammalian host-mediated assay of Legator and cytogenetic tests in vitro and in vivo on animal or human cells. More definitive systems include those based on Drosophila melanogaster, the specific locus test in mice, the dominant lethal in mice or rats, and human epidemiologic studies.

The practical application of these test systems should stimulate further basic and applied studies in relation to such areas as underlying mechanisms of chemically-induced genetic damage and structure-activity investigations on congeners of known mutagens. Realization of the specific limitations of particular test systems may lead to development of approaches for overcoming or minimizing such limitations. For instance, incorporation of mammalian liver homogenates into microbial test systems, or administration of suspect mutagens to mammals with subsequent testing of body fluids in microbial systems might considerably increase the human relevance of data from such systems.

While it is likely that no one test procedure is at present by itself adequate for screening all known mutagenic agents, a combination of the types of procedures mentioned above will probably deduct most classes of mutagens. Such a balanced and integrated approach should also tend to minimize alarmism based on attempted extrapolation to man from test systems of problematic relevance.

In view of the availability of a wide range of, mutagenicity tests, it is surprising that systematic investigations on genetic effects of chemical agents in our environment hardly exist at present. Inadequate attention to this aspect of environmental hazards, with a consequent lack of appro-priately directed research and legislation, represents a major deficiency in public health policies. There is urgent need for large scale screening programs on chemical mutagens already present in the totality of the human environment to include air and water pollutants, pesticides, food additives, and drugs; as a corollary, human exposure to newly synthesized chemicals, particularly those structurally related to known mutagens, should be prevented until such chemicals have been tested for mutagenicity. Such chemicals that are or will be widely used by a large segment of our population of child bearing age must receive particular priority. Particular attention must also be directed to occupational exposure to chemical mutagens, such as ethyleneimines. Screening program should be oriented towards-rapidly developing information to provide a rational basis for legislation analogous to that existing for radiation hazards. Planning should also reflect multidisciplinary approaches and transcend the apparent limitations which have hitherto isolated mutagenesis from other aspects of toxicology. Mutagenicity testing must be incorporated into the general armamentarium of toxicology. It is regrettable that publications on genetic hazards due to chemical mutagens are still rarities in toxicological and public health journals and, in general, are restricted to journals read mainly by geneticists.

Appreciation of such considerations prompted the formation of the Environmental Mutagen Society on January 8, 1969, to encourage interest in and study of mutagens in the human environment, particularly as these may be of concern to public health. The immediate objectives of the society include establishment of a register of chemicals tested for mutagenicity in particular systems and publication of a monograph on methods for testing of chemical mutagens. The society also proposes to publish a newsletter and to act as a resource for information and assistance to government, industry, and interested persons.

The founding initial officers of the society are: President, Alexander Hollaender; Vice-president, Matthew Meselson; Treasurer, M. S. Legator; and Secretary, Samuel S. Epstein. The councillors have been selected to represent a wide range of related interests in addition to genetics, such as toxicology, pharmacology, teratology, carcinogenesis, epidemiology, and public health. Membership is open to scientists and other qualified persons who share the stated purpose of the society and who have relevant knowledge and experience. Membership application forms may be obtained from the Secretary.

The importance of basic and applied mutagenic research in evaluating and limiting our exposure to genetically destructive compounds and of concerning ourselves with practical genetic hygiene cannot be overstated. Our concern must extend not only to the contemporary generation but also to generations yet to come.

5

CHEMICALS WORSE THAN RADIATIONS

In 1980, I joined as a Reader in Environmental Sciences in the School of Energy, Environment and Natural Resources. At that time I was awarded a Research Scheme by the C S I R on Fungicide Genetic Toxicology in plant systems. Our HOD at that time was Prof. A. Gnanam. The research scholars working with were his Nominees. But soon he was obstructing our project work. I requested him for a small place in a huge Botanical Garden under his charge- 5x10 feet area. He refused ! He asked me to discontinue and concentrate on Energy studies. So that a level playing field will be ensured among all the faculty in the school. Ignoring my studies until then. A very novel concept. I had to gradually shift to the new field and simultaneously discontinue my old field. The research work of students were continued with the help of Father Peter of Shembaganur's Sacred Heart College. But pressure to change the field to Energy continued from all colleagues and peers.

First I initiated studies on Energy Plants through an M.Phil student. It was a very successful review thesis appreciated by all. Another student worked on the Calorific Values of plants growing in polluted sites. This was also very successful. The third one was a senior staff member of a College reviewing the environmental effects of energy technologies. This was a great assembly work, but he discontinued and left to his college at Kanyakumari. Then I moved to Bharathiar University as a professor in Environment. I got a research scheme there on the genetic impacts on local biota of polluted ecosystems. Politics were unleashed by the then VC himself forcing me to leave for Pondicherry University. Unfortunately A Gnanam came here as VC and tortured me for 7 years. Then I took the topic of genetic effects of Energy technologies and send it to Hindu. Venkataamani the Agriculture Correspondent appreciated it. Soon it was published and shocked and surprised here and there especially those highly prejudiced against me.

CROW, J.F, 1968

A NIH genetic study report, stated: "There is reason to fear that some chemicals may constitute as important a risk as radiation, possibly a more serious one. Although knowledge of chemical mutagenesis in man is much less certain than that of radiation, a number of chemicals --some with widespread use — are known to induce genetic damage in some organisms. To consider only radiation hazards is to ignore what may be the submerged part of the iceberg." The report further states: "Recent investigations have revealed chemical compounds that are highly mutagenic in experimental organisms, in concentrations that are not toxic and that nave no overt effect on fertility.

The importance of energy is such that the rate of social development is often correlated with energy consumption. Now that the word "development" has become a guiding spirit in every sphere of human activity, more so after the liberalization of economy, the power sector is slated for major leaps. Of these, nuclear energy has been under special favor. But the Koodangulam project, proposed to be set up in southern Tamil Nadu, is under a cloud.

Most types of energy involve technologies to detect, convert, and transport before actual utilization. During these stages, radiation and/or chemicals are released affecting the environment. Among these, the genetic impacts on biota, including human beings, set in motion a chain of events. They produce no warning signals and are irrevocable, some of them compounding with the passage of each generation. These exceedingly sinister impacts are, unfortunately, underestimated.

Chemicals and radiation catch human beings in the ambient and occupational environments, through food chain, usage, consumption, and finally accidents, and modify the genetic material. This results in the initiation of cancers in the same generation and birth defects and genetic diseases in the succeeding generations. It is said in the coming decades that 25 percent of the population will develop cancers and over 10 percent the other two afflictions. The energy sector is also a great contributor. Harmful releases may occur during the production, conversion, or utilization of energy. Data suggest that chemicals actually cause greater and wider harm to human genetic health.

NUCLEAR

Nuclear energy has been feared from the times of Hiroshima and Nagasaki, though the lasting biological damage from radiation was known 20 years earlier, thanks to the pioneering work of Muller. Rather than the trail of death and destruction that radiations left, what became significant was the fate of the survivors. Today, we are faced with a situation of slow exposure to low doses that is suspected to be the result of all nuclear energy establishments, within and outside. Consequently, the genetic material of the exposed people is altered. This exposure may be direct, through occupation, or indirect, through water and herbivorous and/or carnivorous food chain. The Chernobyl disaster could be cited as an example here.

Data on cancer victims among people associated with any aspect of nuclear energy from mining to pollution are considerable. There is hardly any nuclear establishment in the world without this scourge making a preferential visitation.

Sterility among the exposed and birth defects among the children of such exposed parent(s) are well-known. Diseases of genetic origin appearing at different points of age are also documented. Much of our knowledge on gene-based ailments has been a direct consequence of studying radiation effects on biota from the mid-Fifties. Thus, it is understood that mining wastes, coolant waters, fallout of testing, occupational exposure, ambience because of leaks, and food contamination, all carry a genetic hazard to the exposed populations. And what is even sadder and more sinister is that most of the victims are yet to be born.

Though nuclear energy is a comparatively recent source of energy in India, contributing about 2.5 percent of the national requirement, it is slated for an increase from the present 1,250 MW capacity to 10,000 MW in the next 15 years, The present number of units, 10, is therefore expected to triple. What then would be the extent of genetic hazards?

COAL

Now, coal is different. Here, the noxious chemicals mimic radiation in their effects on human populations. Evidence is building up to point out that several processes of conversion and utilization of coal release substances are genotoxic. Thus, during the liquefaction and gasification of coal, pyridines, quinolines, anthracenes, and fluorenes are released contaminating air and water. Therefore, a regular occupational exposure and a slow ecological exposure to these genotoxins has become unavoidable. The burning of coal releases, besides the oxides of nitrogen and sulphur, polycyclic aromatic hydrocarbons, and nitropyrenes, are also highly genotoxic in the experimental systems.

Coal has been the chief source of energy in India, as it is in most parts of the world. The present consumption of 150 million tonnes a year, which is three times more than three decades ago, is likely to double in the next 15 years. Not less than 70 thermal stations are fed with this resource. With about 120 billion tonnes of estimated reserves, coal will continue to be the thurst of our energy efforts for many decades.

PETROLEUM

The processing of petroleum releases wastes, many of which are similar to those from coal and contaminate air and water. Combustion of oil produces, besides the oxides of nitrogen and sulphur, polycyclic aromatic hydrocarbons and nitropyrenes. With its unavoidable use in all forms of transport, urban systems are a veritable sink of genotoxins. These toxins being ambient in the air, and epidemiological studies cannot be taken up. Most hazardous could be the chemicals released during experiments involving fossil oil source.

In India, petroleum and natural gas will continue to be explored, extracted, refined, transported, and used. The former is estimated to be around 2.1 Bbl and the latter 2.0 Btu. Most of the basins earmarked for exploration and extraction are offshore. In all its stages of production, petroleum damages the environment-water and air.

BIOMASS

Combustion of biomass releases, as in the case of fossil fuels, nitropyrenes, especially benzapyrene, a highly dangerous genotoxin. Women doing regular kitchen work are said to inhale daily benzapyrene equivalent to 20 packs of cigarettes. Studies conducted in Hawaii have suggested that gene-based diseases could occur in women this way. It has been confirmed in the studies conducted in some Gujarat villages.

Fuel wood will continue to be used by the poorer sections. Whether the new technologies of wood conversion or efficient direct combustion will alter the compulsions of genotoxin release and exposure can only be surmised at this point of time.

BIOGAS

The problem with biogas is entirely different. The slurry may pose problems through the food chain when reused in agriculture and fisheries. Unless biogas picks up as a source of energy and the slurry is extensively used, the genotoxin problem may not attend this source or energy.

Thus the workers in nuclear, coal, and petroleum industries and cooks in oriental hearths carry enormous genetic risks leading to cancers during their lifetimes and defects and diseases among their offspring for several generations to come. Further, anilines, anthracenes, fluorenes, pyrenes, pyridines, polycyclic aromatic hydrocarbons, quinolines oxides of nitrogen, and sulphur are also released into the ambient urban air, subjecting the population to spells of episodic exposure and to a continuum of low-level ambient exposures. These same and other chemical components may be chemically and metabolically modified in the contaminated water and food-chain. The attending threats to the human genetic system from this source are rather inestimable. It may be appropriate to point out that a 0.5 percent mutational load 20 years ago reached a high 10.5 percent now, cancers excluding.

SCREENING AND TESTING

At this point of time, generation of data on the habitats, other ecological situations, and human populations under stress seem to be the immediate necessity. A survey of the human populations under stress and an epidemiological assay for genetic ailments are needed. The nonhuman biological systems may have to be subjected to in situ monitoring wherever energy prospecting is a current activity. Chemicals released by different energy technologies can be tested for genotoxicity as is done in the West. Investigations of the food- chain, wherever effluents from energy establishments are reused for aqua cultural and agricultural purposes, are needed. Hardly any work is carried out in India on the genetic hazards from energy resources and technologies. An entire field is thus open for research.

Researches on the monazite belt of Kerala, which is contiguous with Koodangulam, started in the early seventies, and yielded some profound initial results. But in the next two decades, nothing really worthwhile emerged and so the Department of Atomic Energy closed its project. Towards the fag end of these unsuccessful efforts, however, came the discovery of a micromutation in the hemoglobin of human blood, the significance of which is not known. Yet this is not enough to damn radiation. On the contrary, the absence of this significant effect can be cited in favor of nuclear energy.

Some of these and other in situ situations were subjects of an Indo-German Binational Conference organized by the Department of Atomic Energy in 1984. With more data, the whole issue was conceptualized by this author in 1990 (Environment and Genome, ed: Mendelsohn and Albertini, Alan Liss & Co.). It was revealed that chemicals were omnipresent genotoxins while radiations were locale-specific occasional/rare possibilities. Chemicals not only remain stable, concentrate, and biomagnify, but also transform to worse species. And all these not merely mimic radiations but outdo them in inflicting a lasting damage.

Given the far wider variety, spread, quantum, and exposure of human populations to chemicals, which could be as genotoxic as radiation, are we justified in branding nuclear energy the villain? While the hazardous chemicals are inseparable from the technologies, radiation occurs only during safety lapses. Can one have a technology-intensive, development-oriented liberal economy, on the one hand, and oppose nuclear energy, on the other hand, when it is the only way to meet new India's energy needs?

6

ENVIRONMENTAL MUTAGENESIS A SUMMARY

This has a strange history. I was at NEHU, Shillong as the Visitor's Nominee on the Selection Committee for appointing Faculty in the school of Environmental Sciences. The School covered not only Environment but also, Geology, Geography, Anthropology. Dr. Mary Taher, a lecturer from the Jamia Milia of Delhi, appeared for the post of Reader. We could not select her. But towards the end she wanted to speak to the committee for a few minutes. She requested everyone there to contribute review articles in our own subjects. She specifically stressed that I should give one – because there are many experts at Delhi and all will give more or less similar ones. She wants something very different. Thus came this article the subject of which is not common place. Some colleagues then, also contributed. Finally a book resulted edited by her and her Husband.

I am not pleased by their product. The array of articles did not cover the Environment satisfactorily, The quality of articles was not uniformly good. Publishing needed much more finesse. Yes. This added to the qualifications of the couple. This is also one of the techniques popular in the post globalized India; to collect articles and become an editor, an author.

REFERENCE: Sharma, CBSR Genotoxic ill-health from the pollution of ecosystems, human occupations, habitations and consumption– Monitoring and Management in India [in] Ecology, Environment and Natural Resource Management [ed: Tahir & Hussain, Jamia Milia Islamia], 7-24, Jawahar Publications New Delhi, 2003.

A. INTRODUCTION

Alma Ata conference a few years ago set a goal for the world Governments-Health for all by 2000 AD- probably by controlling infectious diseases and malnutrition. Whether or not this can be achieved they have ignored the effects of chemicals and radiation on human genetic systems and their consequences, i.e. genotoxicity. The objective of the present paper, therefore, is to focus attention to these health problems by seeking answers to some questions- what is a genetic system? What changes are forced here, by what and how? What are the impacts on society? How did the leaderships measure up to these challenges and finally what next?

Genetic system of an organism consists of genes and chromosomes present in all its cells. They control all life processes and finally transmit the characteristics to the next generation. When changes are affected in the structure of genes (mutagenicity) or chromosomes (clastogenicity) or the number of chromosomes (aneugenicity) cancers, birth defects and heritable diseases occur. Such a genotoxicity in pests, vectors and parasites of human relevance enhance the impacts, multiply costs and complicate problems. The human genetic effects are irreversible, without alarm signals, incurable excepting a few cancers and heritable except cancers. And the heritable diseases multiply with each generation.

The genotoxic impacts are globally on the raise for some decades. The chemicalization of society is also increasing unabated due to consumerism and industrialization i.e. in life styles as well as in environment. Thus the genetic syndromes increased from 3000 to 5000, cancers from 10 to 20% and chemicals from 5 to 10 millions during the last

10 years in the west. The developed countries are therefore making determined bids to manage these problems. As the developing countries also follow the same or similar models -socially and economically, ill-health due to genetic insults from chemicals and radiations stand out rather innocuously ominous. Focusing these issues, hither to unattended, is in the national interest, for at least, a debate may be generated.

B. HISTORICAL ASPECTS

The historical roots of the problem of gene based diseases can be traced to the beginnings of this century. All the three types of genetic effects were discovered in plants and animals during the 1920s. That chemicals can cause all these was established by mid-1940s. During 1960s many alarming facts on cancers and birth defects involving drugs, foods pesticides and natural toxins came to light which lead to the concept of environmental mutagenesis. Rapid strides in human genetic research in 1950s and 1960s correlating clinical symptoms with genetic effects, greatly strengthened this theory. This subject came to be known ever since, as genetic toxicology. Though great strides are made in understanding these poignant problems of human destiny in the developed countries, they remained largely unrecognized in the Indian Public.

C. AGENTS AND EXPOSURES

It is estimated that about 10 million chemicals exist in the world-solids, liquids and gases - 2% of which in daily use to different degrees, in the developed countries. These may be naturally occurring or man made- industrials, dyes, pesticides, drugs, pollutants in the air and water etc. Many of these occur in mixtures. The proportions in which they are present are not known. But it is generally agreed that most chemicals are released without adequate toxicity testing. Testing for genetic toxicity was not included as a requirement until a decade ago. Human populations are exposed to these chemicals singly or in mixtures in several ways- by direct consumption, or through food, water or air, by usage i.e., contacting at home or work place or due to accidents. The quantities taken and the duration of exposure greatly vary. The intake may be through skin or respiratory or digestive systems.

D. EFFECTS ON GENETIC SYSTEMS

At cellular level, these agents bring about three kinds of responses depending upon the targets of reach. Some affect only the segregating mechanism during cell division-spindle, centromere and cytokinesis. This results in the chromosomal number in the offspring cells varying from the parent cells i.e. in the number of linkage groups. The second effect is the causation of lesions in the linear integrity of the chromosomes, during cell division. Consequently, the offspring cells may have one or more chromosomes different from the parent cells in the number and sequences of some genes. Finally and more importantly the very DNA-operational units, genes, may be altered in their basic constitution. The offspring cells therefore will have some genes mutated form their parental forms.

The above three types of offspring cells, if continue to multiply, will soon dominate the tissues. They impart to them and their constituting organs, the new characteristics arising out of the altered gene composition or sequences or number.

Thus the Genotoxic agents affect the cells of organisms in three ways. Firstly by altering the number of chromosomes – the process is called aneugenesis, and the agents, aneugens. Secondly by altering the structure of chromosomes - the process is called clastogenesis and the agents, clastogens. Thirdly by modifying the genes – the process is called mutagenesis and the agents, mutagens. A very large variety of chemicals are known to cause these effects through a variety of mechanisms.

That the agents cause effects on the segregating mechanisms or chromosomes integrity of the dividing cells can be realized by screening the relevant tissues for the end points-the ultimate results within a dividing cell. Frequency of these within a cell gives the intensity of damage along the cell lineages. In case of effects on the individual genes, changes in the phenotypic characters under their control alone would indicate the mutation. Such markers are available almost in all organisms.

E. SOCIAL CONSEQUENCES

Whether the cells are altered genetically through aneugenesis, clastogenesis or mutagenesis, the consequences of transformation to the humans are three. The first is carcinogenesis-unending multiplication of tissues forming tumors. These are not inherited. The second is teratogenesis- the effects are seen in the newborns as birth defects. Some of these arise from an impact on the developing embryo in the mother.

The others are heritable. The third are incurable and heritable diseases. These are manifest at different times in the life span of the individual. Some of these are seen at birth time and therefore are grouped together with teratomas as birth defects.

When most carcinogens were found to be mutagens and non - carcinogens, non-mutagens, it was realized that the few thousand genetic syndromes and the few hundred cancers and birth defects prevalent in society must have been induced by aneugens, clastogens and mutagens ubiquitously existing in our environment since ages. All syndromes do not however constitute miseries. They vary in their pains, durations, manageability, social burdens and economic costs. The total impact on the human society by all shades of genotoxicity may be arguably grater and severer than the other ailments of humankind.

F. ECOLOGICAL CONSEQUENCES

In the non-human organisms also the genetic material is well affected, consequently altering specific characteristic in some portions of the populations. Interactions among these are necessary to set up and maintain balances in the ecosystems. Therefore, changes in them may ultimately upset balances. Some of the characteristics that are known to be affected are pest resistance, host susceptibility, soil fertility, breeding system etc.

G. MONITORING THE EFFECTS

Since humans cannot be tested, methods involving other organisms and human tissues have been developed to identify chemicals with genotoxic properties. Though it cannot be affirmed whether humans also respond to the genotoxins in the same manner as the test systems, many genotoxins were found to similarly act in several others. Consequently it became possible to connect the cancers, birth defects and heritable diseases with environmental agents.

The endpoints of segregational errors, chromosome disintegrity and gene mutations can be monitored in the higher plants, animals, cell cultures and microbes. Among the plant systems, meristems of Allium, Vicia, Pisum and Crepis and the whole plant systems such as Hordeum, Glycine, Tradescantia and Zea are more prominent. Among the animal systems, Drosophila and rodents are more frequently used. Cell cultures of rodent and primate species, including man, are generally favored. Among microbes, Bacillus subtilis, Escherichia coli, Salmonella typhimurium, Saccharomyces cerevisiae and Neurospora crassa are the more widely employed.

H. IMPORTANT ASSAYS

The above indicators systems can suggest whether or not the agents are genotoxic, in a direct way, But this approach is inadequate in view of several complications arising out of tissue metabolism and ecology. To account for these contingencies several types of genotoxic assays are developed.

An epidimeological approach is followed while dealing with the occupational stresses that may cause a genetic effect by looking for genetic end-points and or disease spectra in the susceptible populations. Human excreta are also tested for mutagenicity to find out if mutagens or promutagens are among their daily diet. In situ environment is screened for

Mutagen city either through natural inhabitants such as aquatic animals or terrestrial vegetation or by associating monitoring systems with the habitat. In some assays the possibility of agents undergoing activation or detoxification vis-à-vis genotoxicity, in various tissues, is accounted for. However it was revealed that tissues from individuals with pathological history may respond to stresses differently from the normally healthy ones. Therefore relative propensity of different historical backgrounds to respond to genotoxins has developed into another line of assaying. That in vitro assays should involve the cell types which are the usual targets of the stresses is increasingly realized, e.g. lung tissues for inhaled agents. Finally that lethality may be due to genetic reasons rather than toxicity was fortunately realized right from the early times of this subject.

Though over 100 bioassays are available now, involving microbes, lower and higher animals, mammals and human tissues, for testing the putative genotoxins, only a score of these are of greater value, a few of which are given here

1. **Ames Assay**
   Dr Bruce Ames designed a method in such a way that the chemical (a) gets modified by the mammalian liver enzymes (b) certainly enters the bacterium, Salmonella and © reacts with the bacterial DNA, if capable of. The bacterial characteristics are altered as a consequence. This test revealed that most carcinogens are mutagenic. There is no peer to this assay and is therefore the most employed globally. Of course, several modifications are mode in this to suit specific needs.

2. **Mammals**

Tests involving the genes and chromosomes of plants, insects, mammals and humans- whole body or tissues - reveal all types of genotoxicity. Results from these provided the wisdom to the origins of human syndromes from defective chromosomes. Excessive effects on the genetic system often results in death of embryos. This is monitored in pregnant mammals like rats and mice. Sublethal effects on embryos, i.e., teratogeny are also found here. This is not only the most reliable for linking genotoxicity with teratogenesis, abortions, birth defects, but also the most expensive one, used therefore sparingly.

3. **Human Populations**

Human populations, connected directly or indirectly with pollution- living or working in polluted sites or consuming biota from such sites -are monitored for genetic effects. This is done by (a) medico-social surveys, (b) verifying the health of chromosomes in the blood, amniotic fluid and aborted fetuses and checking the status of male gametes.

4. **Tier or Battery Approach**

For a planned screening of chemicals, generally a group of tests are used following either (a) tier approach or (a) battery approach. The first one consists of a step by step screening, the later steps becoming necessary only if the earlier tests reveal genotoxic potential. The second one involves a simultancous screening in a group of selected tests and assessing the results by a consensus evaluation.

Both have advantages as well as disadvantages. Besides being expensive and time consuming, these exercises encompass the scientist, technician, manufacturer, trader, medical profession, regulatory agencies etc. The vagaries of judging a chemical as a genotoxin or not so as to ban or continue its use, are indeed very complex.

I. MODULATIONS OF EFFECTS

The genotoxicities however are subject to modifications. (1) The genotoxins may be inactivated or innocuous ones may be toxified by the human tissues; (2) the internal effects may be neutralized before manifestation. (3) the manifestation may be prevented from becoming a health risk; (4) The defects may sometimes appear generations later. These are called modulations, enhancing or reducing or nullifying an original effect. These modifications are caused by physical agencies such as temperature, sunlight and pH and chemical processes such as oxidation, reduction, conjugation and a score of others. Further all tissues are not equally sensitive to the toxic properties of chemicals. Such differential behaviors led to the discovery of supermutagenesis and pseudomutagenesis- the former sinister and the later misleading. Deviations from the norm are also caused by excretion of toxins, cellular regeneration, excessive cytotoxicity, barriers to toxin entry, chemical deposition, accumulation, dose-thresholds, excessive gamete production, monthly rejection of some female tissues, abortion of defective embryos and ageing. Some of these increase genetic hazards (deposition, accumulation, activation of chemicals and ageing) while others decrease (detoxification, inactivation, antagonism, excretion, cellular regeneration, barriers, cytotoxicity, male and female gametic dynamics). The latter are predominant.

J. DEVELOPED WORLD

The developed countries are much seized up with the problem of genetic ill-health ever since the foundations of genetic toxicology were laid during the late sixties. Laboratories involving in these problems have gradually increased. Genetic Toxicology became a part of curricula. Environmental Mutagen Societies sprang up in many countries and specialist Journals were launched. So far six international conferences were held in USA, UK, Sweden, Japan and Australia. Both in Europe and USA, the private industry has toxicity monitoring centers. There is at least one country, Hungary, with impeccable genetic health statistics which other countries like Holland wish to emulate. These statistics have given rise to the concept of detriment index - the detriment a genetic syndrome causes to its bearer as a human resource.

Reminiscent of thalidomide teratogenesis three decades ago, a few scores of deformed children are borne recently in the coastal regions of Germany and UK. Similar cases are also found among the seabirds. The inference is obvious. Highest incidences of skin cancer are recorded in Australia and New Zealand (40/100.000). The recent gulf war has already yielded its genotoxic crop of limb deformities and miscarriages. Pesticide connected birth defects were found in Hungary. Electro Magnetic Fields are the current suspects. UV-ray fallout is expected to be high in the regions of 40-50 degrees of latitude The most widespread genotoxic symptom is the reduction of sperm production by 80% and fall in its quality during the last 40 years.

They have now 10 million chemicals, 10-15% cancers and 4-5% birth defects. OECD found that chemicals were never tested for genotoxicity until 1980 and that there are at least 80,000 of them in daily use and about 1000 most frequently. Testing these important ones alone was estimated to cost Rs.55-60000 crores ($ 20 Billion). Toxic Substances Control Act came into existence. Environmental Mutagen Information Centre (EMIC) was organized and a National Toxicology Programme (NTP) was launched in USA where the budget allocations for genetic toxicology increased year by year. International Committee for Protection against Environmental Mutagens and Carcinogens (ICPEMC), International Registry of Potentially Toxic Chemicals (IRPTC) and International Programme for Chemical Safety (IPCS) were initiated.

K. INDIAN SOCIETY

Mutagenic hazards to society from the environmental agents are more or less directly proportional to the level of industrialization and standard of living. In India, a few sources of alarm from this angle can be identified. Chemicals in the industrial atmosphere such as solvents, heavy metal inputs and let offs are an expected source of hazards. Formaldehyde and Cadmium are already known in this respect.

Use of pesticides poses a very great problem, for; they could not only expose the handling personnel but also have an ecological effect by contaminating water and food. A mutagenic pesticide or its metabolite constitutes a very dark threat. The problem here is complicated by the improper use of them despite warnings.

Wastes constitutes another source of possible long term hazard. While sewage is present in all towns and cities, other wastes have reached menacing proportions in about seventeen major cities of India. While these can endanger the water sources, some constitute a new dimension due to their usage in agriculture and other similar cultural practices. From this point of view, one could expect susceptible or stress groups to exist in every township of India.

Source of energy production, conversion and utilization are another area of genotoxin emission and spread. This source pervades from the phytomass use in the rural kitchen to the nuclear source in the Atomic Power Plants. The most common and continuous, however, are the nitropyrenes in the ambient air released due to fossil fuel combustion during transport. Chemicals in the consumable items -tobacco, betel nut and alcohol, constitute a daily source of hazardous substances whenever their production is industrialized which, unfortunately, is ever on the increase.

1. **Life Style Issues**

   Indian habits of betel quid usage- betel leaf, lime, areca nut, tobacco- chewing, snuffing, smoking; rural energy usage- dung cake and wood fuels and some food habits have all genotoxic potentials. Processed foods, with a repertoire of thousands of additives, cooking procedures like deep frying, roasting in direct fire, consumption of pickles, sauses, ice creams, usage of cosmetics and electromagnetic appliances, gastronomical niceties, intake of excessive drugs, liquors and chlorinated water entail exposure to genotoxins such as pyrenes, anthracenes, Polycyclic Aromatic Hydrocarbons, heavy metals, Nitrosamines and pyrolyzed amino and fatty acids.

2. **Pollution Sources**

   Improper usage of pesticides, chemicals in food industry, pharmaceuticals, radioactivity, bye-products from water chlorination, industrial pollutants in air and water, food animals and plants growing in polluted waters and soil have all potential to cause genetic effects in human populations. Work atmosphere in various industries constitute a threat to the human genome. Greater UV-ray fall out in the southern latitudes of India pose a cancer threat. The monazite beach sands of Kerala expose people to beta and gamma radiations. The 50 000 X-ray facilities involving upto a million workers and over-exposed clientele are causing safety concerns to the Atomic Energy Regulation Board (AERB). Kitchen air with traditional combustion and diesel exhausts have Benzpyrene, a very dangerous genotoxin. Several coastlines, riverfronts, urban habitats, Sewage, waste dumps, environs of industrial establishments within and without have similar genotoxic potentials.

3. **Food Chain Hazards**

   As stated above, many toxins enter humans through food chain including pesticides and fungal toxins like aflatoxins. Flavonoids present in the mushrooms and vegetables are convertible into genotoxins in the human system. Some foods may be adulterated and some may contain dangerous contaminants due to inefficient technology of its production. The effluents from various industries contaminate aquatic habitats with PCBs, PAHs, etc, which enter human systems through food chain.

L. PERSPECTIVES AND IMPERATIVES

From the foregoing it would appear that genotoxins and genetic diseases are everywhere. In reality this is averted by the modulations which allow only a fraction of the potential dangers to materialize. Secondly cancers are in far greater frequency than the teratomas and heritable diseases. This is due to barriers for the entry of genotoxins to the reproductive tissues and elimination of affected cells prior to manifestation. As a result, cancers in the west are four times more frequent (20%) than the other genetic defects (5%).In India Cancers are expected to be about 6 million (0.6%) by the turn of century. But with the consumeristic growth in this decade expected to equal the same in the preceding five decades the cancer frequencies will be many times greater than the current projections. The other genetic defects will also increase correspondingly. The net frequency of the genetic pestilence can be anywhere between 1 and 5% and most sufferers are likely to be from the upper and middle classes! So what next? Conceptual changes from the existing mindsets and implementing a programme of action by the leadership classes seem to be the only choice left because hoping to change the development model became a lost cause.

1. **New Paradigms: Detriment Index**

   The 4-5000 genetic syndromes documented may manifest at different stages during the life time of the bearers from birth time to a very late stage of life. Their effects on the bearers also vary from nil to death through minor inconveniences to major suffering. Their frequencies in the populations also vary from extreme low to high. Thus every genotoxic symptom may not inflict suffering on the human and when they do so, not to the same extent. In other words, the detriment the syndrome imposes on its bearers differ greatly. Based on this and with the aid of Hungarian Statistics, an envy of all the developed countries, it was considered that 60% of people suffer some genotoxic detriment though only 4-5% only carry birth defects and 25-30% develop cancers (Shankarnarayanan, 1985 Annual meeting of the EMSI, BARC, Bombay). Thus the human resource is detrimented considerably from insults to their genetic material. The causes of these are the products and bye products of the development. In other words, by following a pathway of consumerism which the west seeks to abandon, refusing to learn lessons from their sufferings and jettisoning native wisdom to care for the future generations the present leaderships are actually planning a detrimental policy and not a development policy.

2. New Paradigms: Gene-Epidemics

The genetic detriments- cancers, birth defects and heritable diseases have spread all over the globe unlike the communicable ones which are essentially local and controllable. Further the genetic systems of unfriendly organisms- pathogens, pests, parasites and vectors- keep changing under stress, thus becoming uncontrollable. On integration, all these constitute genepidemics with enormous costs- personal, social and economic.

7

GENETIC TOXICITY OF PESTICIDES

During early 1987, I was awarded a major scheme by DOEF to study in situ genetic effects from sewage accumulated ponds in Coimbatore. At this time I was the Professor of Environmental Sciences at Bharatiyar University, Coimbatore. There was jubilation everywhere. But my mind was ceased with some sadness that I had to leave pesticide genotoxicity studies. At that time there was onlyone doctoral student, pursuing the pesticide studies.

Why not review the whole subject of Pesticide Genotoxicity as a finale to this branch as we already screened 66 Indian pesticides and developed fully 2-3 plant systems. So we began bringing up this aspect studied in Sweden, Egypt, India, USA, etc. The idea was borne but the priority was to complete the Doctorial Thesis of Panneerselvam.

After he obtained his Ph, D, we wrote to CRC press that we wish to write a review on pesticides. Reply came immediately that they commissioned already another scientist to write on the same subject. Cold water was put on our idea that was put into action already. A week later a letter came from Professor Grant of McGill University, Montreal, Canada inviting me to join him in writing this review ! I agreed immediately. Some weeks later I sent him my drafts. He wrote to the publishers withdrawing himself from this assignment, saying Sharma can continue the assignment successfully alone.

I completed the review "Genotoxicity of pesticide in plant systems". All tables and references were prepared by my fresh doctorate Pannerselvam. I checked them for accuracy and drafted the script, reviewed and dispatched to CRC Critical Reviews, Florida, USA.

The script was critically examined by experts and accepted by another, Prof. Ma of Illinois University, He is the same academic who recommended me for an USEFI fellowship in his university, but torpedoed by the VC of our own University, KVS, in preference to a country cousin with in our school who incidentally left PU after maneuvering maximum benefits. So much for linguistic prejudices and academic moralities.

This review embodies te work of 6 doctorates who worked under my guidance, between 1976 and 1987, covering all the Indian pesticides. This as a summation of my research work during 1975-1987. Happiness around at the success of a unidirectional focus on a single theme and completing it. Credit goes to the patience and perseverance of those 6 students at my idiosyncrasies and limitations! This is some satisfaction absent currently.

This review helped my co-Author Panneerselvam to obtain fellowship in Saville, Spain for 2 years with Prof. Cortes. Similarly he was invited to Germany. Undoubtedly he was the best of the Six. Unfortunately he could not get entry into any University in Tamilnadu. This is because of the tentacles of 9 Octopuses in my department. Not only me, my students also should be punished.! This is our morality and human quality that is ruling Pondichery University for 35 years, so far.

REFERENCE: Sharma, C.B.S.R.and Pannerselvam, N. GeneticToxicology of pesticides in plant systems. Critical Reviews in Plant Sciences. 9: 409-442, 1990, USA.

I. INTRODUCTION

A. Historical Aspects

Since the mid-1940s, synthetic chemicals have become an integral part of modern society. Among these, pesticides are the most notable because they are basically designed to be toxic to man's enemies in the environment, at home and in agriculture, embracing a wide range of target organisms. The use of such chemicals is not new. Legend has it that the Romans destroyed Carathage in 146 B.C. and ploughed the land with salt so that all living things would be destroyed. During 70 A.D, Pliny the Elder recommended arsenic as an insecticide, and the Chinese are, in fact, reported to have used arsenic sulfide as an insecticide in the late 16th century.

B. Pesticide Use

Today, pesticides are considered an essential part of the current agricultural production technology and have been instrumental in reducing losses to crop yields from insects, pathogens, and weeds. They are also indispensable in insect and disease control in forestry, home gardens, lawns, the clearance of roadsides, and the maintenance of right-of-ways. The use of pesticides against the disease-spreading vectors, and hence their indispensability in public health, is also well established.

A brief account of pesticides may be in order. Kurinnyi and Pilinskaya estimated that there are about 800 pesticides. Khilchevskaya estimated that 900 basic chemicals constitute about 100,000 varieties of formulations all over the world. The Pesticide Index of the Tropical Pest Management also lists nearly 800 chemicals, including growth regulators. Ridgeway et al. on the other hand, counted 1500 chemicals, including stimulants, emulsifiers, and solvents, as base chemicals for 50,000 products. The latest estimation of these has reached nearly to 2500.

Concerning the global estimates of the quantity of pesticides used, no reliable data are available. However, in the U.S. until 1984, about 1 billion lbs. of pesticides were used annually for pest control, costing about $2.2 billion, resulting in a gross return of $8.7 billion in the form of crop savings. The crop losses in the rest of the world were expected to be over nearly 10 times than the U.S. The inference that pesticide inputs will be constantly on the rise is inescapable. Although it is believed that the U.S. alone accounts for half of the world's production and consumption, quantities of agricultural usage vary from country to country. Japan consumes 10,000 g, Europe 2000 g, the U.S. 1500 g/ha. India may be cited as representative of developing countries, where more than 100 pesticides have been in use, putting usage at approximately 400 g/ha.

C. Pesticide Effects

Pesticides necessitate close scrutiny, not only because of the large annual volume and varieties of chemicals used, but also because of their high stress index. In addition, they are the only chemicals to which the greatest variety of organisms are exposed, directly or indirectly, sooner or later. Furthermore, the number of humans that come into contact with pesticides during their manufacture and application must be very high and, in fact, immeasurable. The herbicide trifluralin alone is estimated to involve 470,000 applicators and 38,000 field workers in the U.S. In addition to the pesticides *per se*, there are the emulsifiers, solvents, etc. that are combined with pesticides which could be harmful on their own right.

Besides being toxic to the targets as per design, pesticides have other effects on these as well as on nontarget systems. These effects are the genetic effects recognized for their longterm value. The genetic purity built across thousands of years, and therefore needing careful preservation, is currently under threat by the not so careful usage of pesticides. The genetic effects may be genic (i.e., mutagenicity), chromosomal (i.e., clastogenicity), or genomic (i.e., aneugenicity) to various biota in the agroecosystem, including man. These three terms together are referred to as genotoxicity, and the latter two as karyotoxicity. Pesticides, thus, are known to cause heritable defects, cancers, and birth defects in man and contribute to other health factors such as heart ailments, aging, and cataracts. All these obviously threaten the genetic health of current and future generations. The effects on the flora and fauna and on the ecosystems are also well documented.

D. Pesticide Genotoxicity

Considerable emphasis has been placed internationally in studying the genetic hazards of pesticides. Mrak's report gave the first alarm and the rest have followed ceaselessly. It the genotoxicological information on pesticides had grown enormously in scope and depth. Consequently, an unprecedented output of information on the clastogenic, aneugenic, and mutagenic effects on microbes, fungi, plant, insect, mammalian, and human bioassay systems has accumulated. As a natural corollary, 80 to 100 bioassay systems involving the above organisms to screen for environmental chemicals have so far been developed[130] and attempts to enlarge this facility are never waning.

E. Plant Test Systems

Among the test systems, plants appear to be the most important because initially, as well as ultimately, they are the recipients, not withstanding their evolutionary distance from man. Despite this fact and the usage of tobacco and eggplant decades ago to study the effects of fumigation by Nicotene sulfates the plant systems did not gain credibility in genetictoxicology until recently. Among these, root meristems are a favorite material, because of minimum cost and maximum results in scoring karyotoxicity, as a short-term prelude for further studies following the great tradition of Levan's *llium* test. To a lesser extent, the same is true with barley following Nilan's pioneering and standard work. Some of the other plant test systems in use are *Glycine, Nigella, Bellavalia, Haplopappus, Pisum, Tradescantia, Crepis,* and *Arabidopsis.* In fact, any plant under cultural practice and suitable cytogenetic organization can be used to monitor pesticide stresses to its genetic system.

F. Current Status

It is generally regarded that about 20 to 25% of pesticides have been investigated for genotoxicity, half of which may have elicited positive responses in one or the other system. On this provisional accounting, only about 10% of the pesticides appear to have genotoxic potential. Although such a state of our knowledge leaves a considerable number of pesticides, as well as systems, unexplored, the accumulated knowledge is so vast that stock-taking has become necessary. Therefore, summation and evaluation of pesticide genotoxicity in plants comprise a historical need to help keep future investigations on sounder footing and give proper direction since plants are the eventual recipients of pesticides and are the primary sustainers of man.

II. METHODOLOGY

A. Tissues of Exposure, Processes, and Endpoints

Seeds, primary and secondary root meristems, prereproductive shoots, young spikes, and flowers are the organs usually subjected to pesticide exposures by immersion, spraying, or absorption. Following such treatments, root meristems, embryonic shoot meristems, seedlings, meiotic cells, pollen grains, leaves, petals, staminal hairs, pods, or ovules have been assayed for genotoxic effects which are indicated by specific markers depending on the affected process.

The process of mitosis, involving chromosome segregation, chromosome integrity, and cytokinesis, is generally subjected to pesticide stress in any of the growth points, preferentially the root. In meiosis, in addition to the above three, chiasma frequencies may also be stressed. The meiotic impacts may therefore lead to effects on microsporogenesis, megasporogenesis, embryogenesis, and, finally, fertility. Of course, mutations can take place anywhere along the line in both processes.

The consequences of the effects of pesticides on various processes are indicated by changes in specific endpoints. Scoring these endpoints constitutes a sizeable literature in biology Based on their origins, and consequences, the effects may be classified as: (1) pycnotic symptoms or cytotoxicity, (2) chromosome disintegrity or clastogenicity, (3) segregational inaccuracy or aneugenicity, (4) genic changes or mutagenicity. The first three are cytological and the fourth is a morphological event occurring anytime in the plant life cycle. Some endpoints such as fertility could be a cumulative consequence of one or more of the above effects at the cellular level.

1. *Cytotoxicity*

   Cells manifesting effects such as nuclear pycnosis, clumping of mitotic phases, and/or excessive nuclear condensation are not likely to recover to normalcy to complete the division process. Consequently, the other effects caused alongside the above are not transmitted to future cell generations. Therefore, cells with such symptoms should be ignored in scoring for frequencies of aberrations. However, the dosage at which these symptoms appear should serve as the upper limit of experimentation, i.e., the cytotoxic threshold.

2. *Clastogenicity*
   Breakages may be caused in the DNA and at the subchromatid, chromatid, and chromosome levels. While some of these are repaired, others manifest themselves in the successive cell cycles in various forms. Such a loss of integrity is indicated by gaps, fragments, mottles, bridges, and exchanges. Deletions appear as heteromorphic diplochromosomes and micronuclei. These effects, known as clastogenic effects, monitored after long recoveries possess high transmissibility and are therefore of greater genetic consequence

3. *Aneugenicity*
   Inaccuracies can occur in the equitable segregation of the genome into the resultant two cells during cell division. This may occur at two stages. First, if the spindle mechanism responsible for anaphase segregation is affected, the consequences will be seen in the same cell cycle. Metaphases may be totally arrested. Chromosomes may become vagrant, moving either precociously ahead or lagging behind the remainder of the chromosome pack. These could in due course lead to aneuploidy and tetraploidy.

   On the other hand, if the phragmoplast is inhibited, cytokinesis does not take place to delineate the telophase nuclei into daughter cells. Two mitoses will consequently occur in the same cell in the next cycle, which has potential for tetraploidy and/or aneuploidy. Although Brogger 45 initially used the term turbagenicity for spindle errors, it may be better to use the term aneugenicity. Inclusion of the cytokinetic inhibition also into this term may be appropriate because both lead to the same segregational type of inaccuracies. Occurrence of aneugeny long after the chemical stress is removed suggests high transmissibility of the effect and therefore is of greater genetic consequence. Cells with clastogenic and aneugenic manifestations may however be called aberrants in the customary fashion.

4. *Mutagenicity*

Mutations originate during the replication of DNA in the cells of the organ exposed to the toxic agents. The effects are manifest in the next or subsequent generations depending upon the ability of the mutants to cross the developmental and reproductive barriers. However, some would be manifest in the current generation. The specific characters affected may be monogenic or polygenic, morphological or biochemical in nature.

III. EFFECTS ON PLANT SYSTEMS

A. Definitions

1. *Clastogenicity*

Clastogenicity means lesions in the longitudinal integrity of the chromosomes and their resultant manifestations, as ex-plained in an earlier section. Gaps, fragments, bridges, micronuclei, exchanges (subchromatid, sister chromatid, chromatid, and chromosomal), heteromorphic diplochromosomes, and stickiness in the mitotic and meiotic cells indicate such lesions. All these are likely to result, on transmission, in an altered sequencing in the genetic linkage groups.

2. *Aneugenicity*

   Aneugenicity means disturbance in the segregation of chromosomes during cell division which was explained in an earlier section. Lagging and precocious chromosome movement, metaphase arrest, micronuclei, and multipolar segregation due to split-spindle are due to spindle dysfunction. Cytokinesis dysfunction results in binucleation and bimitosis. In meiotic cells, in addition to the above, noncongression, and nondisjunction and delayed disjunction of bivalents also indicate aneugenicity. All these are likely to result, on transmission, in an altered genome.

3. *Mutagenicity*

   Mutagenicity endpoints assayed are comparatively very few such as petal and staminal hair color *(Tradescantia)*, pollen waxiness *(Zea)*, laminar pigmentation *(Glycine)*, embryo lethality and morphological characters *(Arabidopsis)*, and chlorophyll deficiencies *(Hordeum)*. Since they are mutations, such effects will result in the altered composition of the gene pool.

B. Effects in Various Organ Systems

1. *Mitosis in the Root Meristems*

Primary and secondary root meristems of a number of higher plants are used extensively for assaying the pesticide effects on the spindle, chromosomes, and cytokinesis. As is shown later, they are also the most improperly used among the assay systems. These organs are, however, developmentally inconsequential, except perhaps in the adventitious systems. Being the quickest to yield data on cellular endpoints, these are best put to use as short-term prescreen tools, especially where finances and facilities are meager. Nevertheless, the data on the effects of meristems have been unequivocally accepted. Of the 150 pesticides investigated for their effects on the mitosis of the root systems of various plant systems, 40 were fungicides, 60 herbicides, and 50 insecticides. Of these, 90 were positive in their impact and 10 negative; 20 elicited dose responses and 40 did not. In the experimental design, the treatments given were both chronic and acute. Recoveries — nutrient or aqueous — were rarely carried out. Controls — negative and/or solvent controls — were also rarely maintained, while positive controls were not even mentioned!

Of the 40 fungicides tested, 20 had effects on mitosis and 10 did not; 10 exhibited dose-response behavior and 10 did not. The effects and behavior of the remaining compounds are uncertain, weak, or questionable. The most effective fungicide appears to be benomyl. Of the 60 herbicides tested, 30 had effects on the mitosis and 5 did not. Only five exhibited dose- response behavior and ten did not. The effects and behavior of the remaining compounds are uncertain, weak, or questionable. Simazine, paraquat, maleic hydrazide, 2,4-D, and atrazine appear to be the most investigated. Of the 50 insecticides tested, 30 had effects on the mitosis and 5 did not; 15 compounds exhibited dose-response behavior and 15 did not. The effects and behavior of the remainder are uncertain, weak, or questionable. Dichlorvos, lindane, malathion, parathion-methyl, and phosphamidon appear to be the most investigated. All pesticide counts mentioned above are approximate.

2. *Mitosis in the Embryonic Shoot Meristems*

Shoot meristems provide a valuable tool in genotoxicity studies because, even if the error caused in these meristematic cells is slight, it has the potential to result in many mutant cells in the adult plant. The aberrant cells in this tissue are significant for two reasons: (1) their biochemical origin in the preceding cell cycle and (2) their transmissibility to the adult plant because the short meristem is but a system enroute to the germline. The transmissibility of effects from shoot meristem to germline, from mitosis or meiosis, is subject to diplontic selection. In other words, if the floral primordial cells of the embryonic shoot are affected, karyotoxicity will manifest itself in the sporogenous tissue. There-fore, this system is unique in genetic toxicology.

In the experimental design, the treatments were acute, targeting the "S" phase in the seeds, followed by metabolic recovery rendering the systems highly reliable. Both positive and solvent controls were maintained.

Of the 45 pesticides tested for their effects on the mitosis of the shoot meristems, 20 were fungicides, 15 herbicides, and 10 insecticides. Interestingly, all this work was carried out in a single system, barley, which was very well defined. Of these 45, 25 were positive and 15 negative in their effects; 20 showed dose-response behavior and 5 did not. The effects and behavior of the remaining compounds are uncertain or weak.

Of the 20 fungicides tested, 10 affected the mitosis in the shoot tips and 5 did not; 10 exhibited dose-response behavior and 5 did not.

The most effective fungicides appear to be anilazine, captan, ethirimol, and zineb. Of the ten herbicides, five are positive and five negative in their effects on the shoot tip mitosis. Only three of these manifested dose-response behavior. The most effective appear to be chlorpropham and monolinuron. Of the ten insecticides, only five affected the mitosis and all these demonstrated dose-response behavior. The most effective appear to be oxydemeton methyl, parathion methyl, and phosphamidon.

All the above pesticide counts are approximate.

3. *Meiosis in the Male Germline*

Pollen mother cells offer a better system to score the clastogenic and aneugenic abnormalities than the embryonic shoot cells because of their natural synchronous activity. Some end-points, like fragments, bridges, micronuclei, and vagrancy, are common in mitotic tissues. In addition, delayed disjunction, nondisjunction, noncongression, and lagging of bivalents occur in the pollen mother cells. The aberrent meiosis, apart from causing structural changes in chromosomes, may lead to aneuploidy in the next generation.

Of the 90 pesticides so far tested for effects on the meiosis of the male germline, 30 were fungicides, 30 herbicides, and 30 insecticides. Of these 90, 60 pesticides affected the meiotic process, mostly tested in the barley system, the others being *Vicia* and *Tradescantia*. Among these, 25 showed dose-response behavior and 15 did not in barley. The effects and behavior of the remaining are weak or uncertain.

In the experimental design, the treatments were generally acute, followed by a very long recovery, i.e., from seed to flower, or a short recovery, i.e., spraying on and assaying the flower buds. Positive as well as solvent controls were generally maintained, giving credence to the work.

Of the 30 fungicides tested, 15 affected meiosis, only 10 of which demonstrated a dose-response behavior. Anilazine, cap-tan, ethirimol, and zineb appear to be the most toxic. Of the 30 herbicides tested, only 15 affected meiosis, only 5 of which demonstrated dose-response behavior. Chlorbromuron, chlorpropham, and monolinuron appear the most toxic, although most work was done on atrazine and 2,4-D. Of the 30 insecticides tested, 20 affected meiosis, of which 10 demonstrated a dose-response pattern. Ethylenedibromide, ethylenedichloride,

BHC, carbaryl, malathion, and phosphamidon are among those most effective.

4. *Chiasma Frequencies in the Meiocytes*
It is well known that chiasma frequency is not only a sensitive and scorable endpoint, but it also carries a wide range of implications, being a cytological indicator of genetic recombination. While the causes for chiasma depression are laid during the DNA synthetic activity in the seed, as far as the barley assay is concerned how this actually ensues in the pollen mother cells is obscure. However, it is realized that irregular chiasma functioning can lead to aneuploidy in the next generation.

Of the 50 pesticides tested so far for effects on chiasma frequencies, 15 were fungicides, 15 herbicides, and 20 insecticides. All this work was done only in the barley system. There was not a single compound which increased the chiasma frequency. While 30 pesticides decreased the chiasma frequencies, only 20 of these have shown dose-response patterns.

The experimental design, treatments, and scoring are identical to those followed for studying meiotic effects.

Of the 20 fungicides tested, 10 reduced the chiasma frequencies, most of which elicited dose responses. Ceresan, edifenphos, and mancozeb appear to be more effective. Of the 15 herbicides tested, 10 reduced the chiasma frequencies, 5 of which elicited dose responses, with anilazine and diuron being the most effective. Of the 20 insecticides tested, 10 reduced the chiasma frequencies, most of which elicited dose responses. Dichlorvos, edifenphos, and phosphamidon appear to be more effective.

5. ***Seedset in the First Spikes***

Even if the meiotic abnormalities in the male germline lead to some pollen sterility, it would be materially ineffective to reduce seedset due to the prodigality of pollen production. If, however, similar events take place in the female germline, seedset would be affected. This endpoint, reduction in seedset, is therefore considered suggestive of genotoxicity in the female germline.

Of the 30 pesticides so far screened for effects on the seedset, 15 were fungicides, 10 herbicides, and 5 insecticides. Of these 32, 25 reduced the seedset. Among these, 17 showed dose-response behavior. None increased the seed output. All the work on this aspect came from a single laboratory using the barley system.

The experimental design and treatment are identical to those followed for screening meiotic effects. The scoring, however, had to be limited to only the first spike, which alone might have originated from the embryonic shoot meristem — the recipient of the initial pesticide stress.

Of the 15 fungicides tested, 10 reduced the seedset, 5 of which exhibited a dose-response pattern. Ceresan, echlomezol, edifenphos, and mancozeb appear the most effective. All of the ten herbicides tested reduced seedset, most of which eliciting dose-response behavior, with ametryne being the most effective.

Of the six insecticides tested, only two reduced the seedset, which followed a dose-response pattern.

6. *Mutagenicity*

Broadly speaking, mutations are of two types. The first type is a change in specific characters such as staminal hair color, laminar spots, pollen waxy content, chlorophyll deficiencies in seedlings, etc. The second type are the lethals. Coleoptile elongation without leaf emergence and a curved lamina which often fasciates form the lethal types in barley. In others, either the embryos are aborted or the seeds are deformed. Both of these are well documented in the literature covering *Hordeum, Tradescantia, Zea, Glycine,* and *Arabidopsis.*

Of the 85 pesticides so far screened for mutagenicity, 25 were fungicides, 30 herbicides, and 30 insecticides. Of these, 60 were positive. No attempt was made to study the dose-response behavior of the compounds. Also, no detailed studies were done beyond screening for the mutagenic potential at the highest concentration. On this aspect, most of the work came from the barley system .

The experimental design and treatment are identical to those followed for screening meiotic effects, with a single difference. Sometimes only a single highest dosage was taken into account. The scoring for mutations was normally carried out in the relevant organs of the M_2 generation from the M_1 seedset of the treated generation.

Of the 25 fungicides screened, 10 had mutagenic effects of various frequencies. Almost all the 30 herbicides had similar effects. Among the insecticides, 20 out of 30 were found mutagenic. These mutations were chlorophyll deficiencies, lethals, and pollen waxy changes

IV. EVALUATION OF EFFECTS

The stress effects of pesticides on various endpoints in the plant life cycle need not necessarily be inherited. Some effects, such as those on germination, seedling growth, and mitotic indices in roots and shoots, may just affect only the initial development of the plants. Meiotic effects, including chiasmata frequencies, may end up causing pollen sterility and reduction in seedset, i.e., affecting reproduction. However, it cannot be ruled out that clastogeny and aneugeny in shoot meristems and pollen mother cells may indeed change karyotype integrity in the next generation. Mutations, however, are definitely indicative of heritable changes. The pesticides that have effects on various endpoints have the potential to affect development and reproduction besides causing heritable changes. A brief examination may be appropriate

A. Fungicides

Of about 40 fungicides tested so far, about 17 (42%) have shown the potential to cause varying degrees of genetic damage to different aspects of plant life. Of these, 13 were found to interfere preferentially with the processes related to development, 16 with reproduction, and 10 with heritability of characters. Only five — carboxin, oxycarboxin, fuberidazole, thiabendazole, and thiophanatemethyl — do not possess any such effects. However, no conclusions can be made on 18 fungicides due to a paucity of data. Anilazine appears to be singularly the most hazardous pesticide.

B. Herbicides

Among the 71 herbicides screened so far, only 24 (34%) demonstrated the potential to cause genetic damage to various degrees. Of these, 41 interfered with the processes connected with development, 14 with reproduction, and 18 with heritability of characters. Only terbacil has no impact. However, as many as 46 herbicides remain to be adequately tested. Atrazine alone is effective on all the endpoints checked so far.

C. Insecticides

Of the 56 insecticides tested so far, only 16 (30%) were conclusively effective on plant systems. Of these, 11 interfered with processes concerning development, 8 with reproduction, and 13 with heritability of characters. Only formathion had no effect at all. However, as many as 30 compounds remain to be tested adequately. Dichlorvos, fensulfothion, parathion-methyl, and phosphamidon appear to be the most toxic.

Out of nearly 170 pesticides screened with plant systems, only 60 (30%) may be considered genotoxic to the nontarget plant systems for whose protection against pathogens, pests, and weedy competitors these compounds were originally designed. While only ten (6%) can be unequivocally cleared of any such hazards, it has to be stressed that the bulk of pesticides remain to be adequately investigated. Thus, these hazardous pesticides may affect not only the nontarget crop systems, but also may develop in the targets themselves, at sublethal doses, resistance to their pesticides. This aspect is considered subsequently.

V. PERTURBATION OF THE AGROECOSYSTEMS

If the various criteria mentioned in the previous section are applied, the number of pesticides actually genotoxic could be far less than claimed. Prior to such a check, a few considerations may be relevant in the context of current ethos in pesticide testing with specific endpoints in higher plants. They are (1) what happens to other aspects of crop life, (2) how do other biota in the agroecosystems react to the pesticides around, (3) what are the implications of the present knowledge to the whole agroecosystem, (4) what is the current scenario in the agroecosystems under pesticide stress in the developing countries, where the usage is ever increasing, and finally (5) what could be the future ecological portents in the agroecosystems due to genotoxic pesticides?

A. Effects on Crop Systems

From a genetic point of view, mitosis in the shoot meristems, meiosis, seedset, and mutations are the other aspects of importance besides the usually tested meristems of the root. We have seen that all these are subject to pesticide stress. In a comparison, effects on meristems and effects on other endpoints seem to have commonality to varying degrees. A careful perusal of data reveals that 90% of the pesticides producing positive effects on meristems also cause the same effects on meiocytes, 60% reduce seedset, and 20% cause mutations. Obviously, the compounds affecting the meristems will have similar effects on other endpoints too, albeit to reduced extents. Theoretically, impacts on all these endpoints, by themselves, may have some consequences. The effects on meristems may cause developmental variation. Meiotic aberrations may lead to sterility and affect recombination and segregation of characters. Also, a reduction of seedset per inflorescence, being the biological basis of grain yields, may ultimately result in the reduction of crop producitivity. Mutations, whether or not scorable, may involve serious characters, including development of susceptibilities to newer strains of pests and pathogens.

B. Effects on Other Biota in the Agroecosystem

The karyotoxicity and mutagenicity of pesticides were also studied in a wide variety of systems. From the perusal of these data emerges a picture of concordances between plant responses, on one hand, and those of microbes, fungi, insects, and mammals *in vivo* and *in vitro*, on the other. Therefore, the likely similarity of responses between plants and other inhabitants of ecosystem is a cause for concern. A few examples are detailed.

Among the fungicides, benomyl, captafol, captan, carbendazim, fenaminosulf, and mancozeb are genotoxic in plants, microbes, fungi, insects, and mammals notwithstanding a few exceptions). Among the herbicides, atrazine and paraquat are similarly genotoxic in plants and fungi. Among the insecticides, dichlorvos, malathion, methyl parathion, oxydemeton methyl, phosphamidon, and trichlorfon are genotoxic in plants, microbes, insects, and mammals.

These 14 pesticides are therefore the most hazardous to the agroecosystem from a genetic point of view in that they have the potential to subject not only crops, but also pathogens, pests, and others to mutational stresses. The ultimate consequences of these mutational loads to the whole system can only be surmised at this point of time.

Since pesticides are mostly used in agriculture, their genotoxicity to a wider variety of biota may have implications to the stability of the agroecosystem. The system's stability depends upon healthy interactions within the community; interactions depend on the characteristics of the populations of organisms; and the characters are genetically controlled in the individuals. It is these genetic components, not only of the crop but also of the weed, the pest, and the pathogen and therefore of every other member of the biota, that pesticides threaten to affect.

C. Implications to the Whole Agroecosystem

At this juncture, it may be appropriate to ponder the likely targets of mutational stress in the agroecosystem.

It has been estimated that out of 45 fungicides, 80 herbicides, and 60 insecticides screened in plant systems so far, only 25 fungicides, 20 herbicides, and 15 insecticides have been screened for genotoxicity in other biotic systems as well. It would also mean, conversely, that many pesticides screened in other systems might not have been tested in plants.

Among the 25 fungicides screened in various biota, only 12 possess mutagenic potential in both fungi and higher plants. In other words, the crops receiving these 12 fungicides may develop mutations for susceptibility to new pathogens. Similarly, pathogens may develop mutations for resistance to the very fungicides used to control them. In the realities of the agroecosystem, this would mean that pathogenic fungi and their host crops may have to be under surveillance for genetic changes whenever these 12 fungicides are in use.

Among the 20 herbicides screened in various biota, all possess the potential to be mutagenic in higher plants. In other words, the weeds receiving these herbicides may develop mutations for resistance to the same herbicides used to control them. In the realities of the agroecosystems, this would mean that several weeds and the crops are gradually increasing their genetic variability.

Among the 15 insecticides screened in various biota, only 4 possess the potential to be mutagenic both in insects and higher plants. In other words, the crops receiving these four pesticides may develop mutations for susceptibility to new pests. Similarly, the pests may develop resistance to the very same insecticides used to control them. In the agroecosystem, therefore, one should be looking for newer pest strains and more susceptible crops.

In the long run., crops may develop susceptibility to new pathogens and pests. The pests may develop resistance to pesticides due to their genotoxicity; such compounds, by the present reckoning, only number 16 out of nearly 200 screened for genotoxicity. Even this is only an empirical inference - a whole field of scientific investigation awaits evaluation.

D. Agroecosystem Realities - the Current Scenario

It has been well established that the agroecosystem, consisting of crop, weed, microbe, insect, fungi, invertebrates, vertebrates, mammals, and humans, functions in a balanced and productive way due to the conservation of genetic material and stability of each of these components. Recently, there have been reports of the erratic behavior of these components, although pesticide-implicated ecological imbalances have been known for some time· For example, ploidy variations, chromosomal aberrations, and gene mutations were also known to occur in some of the biota relevant to the agro-ecosystem. Furthermore, insect and other pests developing resistance to pesticides through genetic alterations, human workers showing chromosomal changes, and innocuous pesticides often becoming genetically hazardous have been on the increase. This information and other historical and circumstantial evidence led Sharma to propose the theory of agroecosystem instability due to pesticides.

E. The Ecological Portents – the Instability Theory

This theory emphasizes that pesticides mutate various biota in the agroecosystem directly or indirectly, suddenly or gradually. Since crops are the initial as well as the ultimate recipients of pesticides, it is inconceivable that the biota ecologically associated with them could escape the mutational stress of pesticides. It is further postulated that pesticide mutagenicity could alter the nitrogen-fixing capabilities of bacteria and blue-green algae. Crops may suddenly become sensitive to traditionally minor pests. Insects could become resistant to pesticides. Habits and habitats of useful organisms may be altered; breeding systems may be modified. Since many pesticides may stay in the system as residues in the same or modified form, there could be more severe long-term impacts in the food chain.

All these together have the ingredients to threaten the stability of the agroecosystem, which is the very basis of crop producitivity.

VI. RELEVANCE IN GENOTOXICITY TESTING

A. Background

In genetic toxicology, testing chemicals for mutagenicity, carcinogenicity, and teratogenicity and applying the results to humans forms the basic ethos. Since the human species cannot be directly subjected to testing, mammalian species form the all important link between the experiments and the ethos. The question now arising is how far are the results of the plant assays applicable to mammals? In other words, do these two systems respond similarly to the pesticides. If the answer is "YES", the credibility of plants as mainstream test systems is assured. If the answer is "NO", the utility of the systems would be restricted to the agroecosystem only.

Kihlman first attempted this comparison between meristem and mammalian cell responses. Subsequently, Grant' made more successful attempts in this direction. Almost at the same time, a workshop at NIEHS, Research Triangle Park, U.S., found a 99% concordance between selected plant and mammalian systems.' Subsequently, Sharma found a 73% concordance for genotoxicity and 75% for karyotoxicity between these two systems.

B. Concordances

In the present assessment, 25 pesticides have relevant literatures for comparison, i.e., 25 were tested in the mammalian and plant systems. Of these, 15 pesticides elicited similar karyotoxic responses in both. This works out to a concordance level of 60%. Mutagenicity could not be compared because the plant mutant indicators such as chlorophyll deficiency, etc. are morphological endpoints some of which are complex polygenic systems while the mammalian and human indicators were mostly *in vitro* monogenic biochemical markers.

VII. CONCLUSIONS

A. Specificity vs. Nonspecificity

Most pesticides appear nonspecific in their activity at the cellular level. In other words, only very few can be definitely rated as aneugenic, clastogenic, or mutagenic. Even after all the checks have been followed, a considerable number of pesticides may be found to cause more than one effect. However, some specificity is found with regard to phragmoplast inhibition leading to bimitosis. Only 3 pesticides out of about 40 tested so far are known to cause bimitosis. The reason for this situation is the lack of investigations to correlate structures and properties of pesticides with activities.

B. Effective Doses

Similar karyotoxic effects may be produced by varying ranges of doses of pesticides. Some pesticides are effective at very low doses while others are ineffective even at very high doses. These variations could be due to several reasons. Factors such as the chemical nature, permeability metabolism, mechanism of action, and persistence of the pesticides and their impacts on cell cycle, and partial or full repair of the damage during recovery regimen could influence the causation, appearance, and frequencies of effects produced in a system vis-a-vis the dose administered. The minimum effective doses are also likely to be far lower than those obtained hitherto because (1) most of the work was not done in a reproducible manner, (2) many workers used commercial samples which contain many accessories to the pure technical grades of pesticides, and (3) some of the effects may be secondarily caused by pesticide metabolites whenever the interval between treatment and assaying is long, depending pon their half-life periods.

C. Field Applications and Genotoxicity Rating

Despite the innumerable studies on the genotoxic responses of plants to pesticides, the role of field dosage and mode of application were never considered while assessing a compound for genotoxicity potential in the field. The experimentally determined genotoxic dose ranges of the pesticides should be checked whether or not they fall within their prescribed field application dosages. The inference is that only such pesticides are likely to cause genetic effects on crops standing in the field, in due course of time.

Another limiting factor regarding the relevance of experimental results to the field is the mode of application. For example, seed dressings may have more initial effects on the root systems and embryonic shoots which may be subsequently passed on to the adult plant. Soil application poses the problem of low-dose chronic impact on all soil-borne biota. On the other hand, foliage sprays cause the adult plants to retain the pesticide and translocate it to the flowering shoots before getting washed down to the soil. While all three modes have an impact on nontarget aspects, seed dressing and soil application appear more consequential at the level of development, while spraying has greater impact at the level of reproduction and producitivity. It would appear therefore from the present survey that many of the positive effects in the experimental systems may not hold true in field situations.

D. The Pesticide Situation in Developing Countries

Pesticide inputs into the agriculture in developing countries has been on the increase and this trend is expected to continue. Yet only a fraction of the arable land is covered by these chemicals. The quantities as well as varieties of compounds per unit land is far less in developing countries compared with the developed nations. However, within this small fraction used, the input is great because of overdosing generated by the "overkill" syndrome of the farming community. In addition, adulteration of the pesticides with inks, kerosene, varnishes, dyes, etc. is also rampant. In developing countries, the genetic effects are thus limited only to that fraction of land (e.g., about 10% in India) where the crop systems are exposed to pesticide usage and the perversions caused. Another major problem pertains to the misuse of pesticides in protecting consumer production, i.e., lacing the containers of grains, flour, and other similar consumables with pesticides — a prohibited practice - as a marketing method. Obviously, the problems of genetic toxicology of pesticides are more serious in developing countries than in the developed world in spite of a higher dosing norm.

E. Effects on the Female Germline

Thus far, there has been no work on the effects of pesticides on the meiosis in the female germline due to the obvious logistic difficulties in handling megasporogenesis. An indirect indication is however offered by the reduction of seedset, even when there is no pollen sterility under pesticide stress. It may be noted that studies on female meiosis itself are generally meager.

F. Indirect Effects

Slow and gradual effects can indirectly occur in the plant systems from three sources. First, pesticides may be metabolically modified in the plant system into genotoxins which may exert influence that the parent pesticide did not possess. Several herbicides are known to behave in this way, releasing yet unidentified mutagens into the host plant. Second, pesticide residues, harbored in miniscule quantities, may slowly affect genetic material of the crop. However, such concentrations at the residue level have not yet been demonstrated to have any genetic effect even in the meristems, the simplest of plant systems. This possibility, however, cannot be ruled out. The third source is limited only to developing countries where grains are often stored in both the farm houses and market places under heavy pesticide lacings and sprays. This improper practice subjects the seed stocks to unusual pesticide stresses which may be revealed when the seeds are germinated to raise a new crop. Many so-called spontaneous variations in crop plants may, in fact, be due to this source. Hardly any work has been carried out on such effects.

8

AGRO ECOSYSTEM INSTABILITY

During 1980-1, I shifted from the position of Reader in Botany in Berhampur University to the position of Reader in Environmental Sciences in Madurai Kamaraj University. as some backlog in research responsibilities still remained to be completed, Students used to visit me at Madurai. During discussions with one such student who came to finalize his doctoral thesis, an M. Phil. student of our department happened to be there. He suddenly and curiously enquired if we are interested in pesticides. In affirmation I told him of our 3 earlier theses on some aspects of pesticides. He was Mr. Manavalan.

Manavalan came next day with a few copies of PPST Bulletin, published from Britain. it was entirely devoted to pesticides and pests. I also became a member of this group. My discussions with Manavalan gravitated from genetic consequences to non-target humans to similar effects to target pests and pathogens. A new dimension is added to our hither to static and limited interest in genetic aspects. in one of the issues of PPST Bulletin, i found an advertisement for papers to be presented in an international conference in Thailand. I prepared a review paper and sent. and Prof. Tait immediately accepted it in principle, but asked me to focus, and abridge. The paper went to her revision 2-4 times. the version presented here is the final version. when I wrote her my inability to attend the conference, an invitation to attend and a grant to meet all expenses came in response.

Thus I went to Thailand and participated in the symposium at Chiangmai, stayed in Bangkok for another 3 days, and returned happily, with renewed confidence and widened vision. This review paper was published in a book edited by Tait and Napompeth and published by West View Press.

In this small but significant review, i also appended a flow chart that was excluded in the final version of the paper.

REFERENCE: *Sharma, C.B.S.R. 1987. Agroecosystem Instability due to pesticides (in) Integrated pest management (ed. Tait & Napompeth), West View Press USA: 92-96. 1987.*

INTRODUCTION

It is estimated that over four million chemicals exist, with a few thousand added annually to which man and other biota are exposed in the workplace, as residues in food or as general environmental contaminants. Toxicity to genetic material is a potential effect of these chemicals and may occur at the gene, chromosome and/or genome level. resulting in mutagenicity, clastogenicity or turbagenicity, together constituting genotoxicity. In man this leads to carcinogenesis, teratogenesis and heritable diseases, adding to the social burden

It is proposed here that the genotoxicity of pesticides is a potential threat to the stability of agroecosystems causing mutations, directly or indirectly, gradually or suddenly and such pollution by imperceptible mutations may alter the delicate balance of the ecosystem.

EVIDENCE FOR GENOTOXICITY

Kurinnyi and Pilinskaya estimated that there were 800 individual pesticides, only 17% of which had been investigated and of these 70% were genotoxic in one or more systems. Ridgeway et al. estimated that there were 50,000 formulations of 1,500 basic chemicals. Khilchevskaya assessed the former at 100,000 and the latter at 900, claiming that 120 pesticides, out of 240 studied were genotoxic. Considerable evidence has accumulated on the genetic effects of pesticides on various organisms, Approximately 50% of all studies were on plant systems.

DIRECT EVIDENCES

In studying components of agroecosystems. ploidy variations, mutations and cytogenetic abnormalities have been shown to occur in a variety of grain crops. Fertility losses in sorghum have also been recorded. Chlorophyll deficient mutations have been reported in barley used as an experimental system. Resistance to herbicides has occurred in both crop plants and weeds. Weeds with genotoxic manifestations have also been recorded from fallow lands with a history of herbicide application.

Genetic changes in rodents in Vietnam have been attributed to herbicide sprays. United Nations Environment Program (UNEP) has recorded seven species of rodent resistant to rodenticides, including *Rattus rattus* and *R. norvegicus*. The Food and Agriculture Organization estimated 182 strains of insect and mite to be resistant to pesticides in 1965, increasing to 364 by 1979. Approximately 223 agricultural pests have developed resistance to nine major groups of pesticides. Cotton leaf worm developed resistance to ten insecticides during 1961-1976 in Egypt. The World Health Organization has estimated that the number of resistant mosquitoes and other arthropods increased from 102 to 121 during 1968-1976. Resistant plant pathogens were negligible in 1965, but increased to 35 species after the introduction of systemic fungicides and the acceleration of herbicide use. The genetic bases of this resistance are well established.

A high incidence of mutations among organisms lower on the evolutionary scale is indicated by a review of pesticide effects on microbes, although literature specifically related to mutations is very meagre.

In India, following the green revolution, several pests increased in importance. A minor disease, Karnal Bunt, became a major one on wheat; yellow and brown rust assumed epidemic proportions in the high yielding varieties of wheat; gall midges, tungro virus, bacterial blight and brown plant hopper became major pests of rice. Possible explanations are that pests are evolving new strains or that the crop varieties are losing their resistance, under the influence of chemical applications, both factors being under genetic control and therefore inherited.

INDIRECT AND CIRCUMSTANTIAL EVIDENCE

The possibility that microbes and mammals respond to pesticides in the same way as do plants is supported by concordance estimation from the published literature. Nearly 80% concordance has been found between plants on the one hand and microbes and mammals on the other. Allowing for differences in rates of exposure, this implies that biota in the agroecosystem may be under genetic stress from pesticides to the same extent as experimental plant systems.

Natural vegetation in habitats receiving industrial effluents showed genotoxicity. Similar results were obtained when tester strains were used as monitors. Genetic abnormalities have also been found in pesticide workers). In Handigodu, Karnataka State, India, about 600 people have been affected by a variety of bone degenerations during the last eight years. The effects are found even among infants and include dwarfism. Consumption of crabs and fish from paddy fields receiving insecticides is epidemiologically linked to these disorders. Similar events on a lesser scale have occurred in other parts of Karnataka State.

CONCLUSIONS

Slow and imperceptible changes may occur in the biota of an agroecosystem under pesticide stress. Pesticide inputs are low in those tropical ecosystems which are most rich in biota, but their use is increasing. Indications of the breakdown of agro-ecosystems may be; crops are becoming more susceptible to pests; pests are developing resistance to pesticides. Chemical interactions between pesticides and fertilizers may lead to new and more toxic compounds. Natural fertility of the soil due to microbes may be depleted. The breeding system of the crop may be modified. Habits and habitats of useful organisms may be altered, affecting host-parasite predator-prey relationships.

Various factors could either accelerate or retard the development of genetic stress from pesticides and knowledge of these moderating factors is fragmentary. Overdosing with pesticides is a potential accelerating influence, while pesticide adulteration diminishes effectiveness against the target species, leading to the evolution of new strains. Many pesticides contain noxious impurities. If genotoxic pesticides are residual in nature they pose a long term hazard in food chains, to an extent dependent on the chemical's half life. Plants may also activate nongenotoxic pesticides into mutagens. Degradation generally removes the hazard, but can generate more dangerous compounds as in the case of diflubenzuron and 1.2-dibromoethane. Diluents, adhesives, adjuents and wetting agents are generally considered genetically inactive.

Developing a proper perspective on imperceptible mutation pollution by pesticides in agroecosystems, with ramifications through the food chain, is a necessary part of environmental monitoring for pesticide hazards.

9

GENE-EPIDEMICS FROM PESTICIDES

I took lien from Berhampur University and joined Madurai University. I came here for an Interview for the post of Reader in Plant Genetics. The HOD of Botany did not like me. But all others in the selection committee, including VC did not agree with him. They send me an appointment letter as Reader in Environmental Science based on my work on pesticides. I arrived and joined. I had then a C.S.I.R. scheme on fungicides with two vacancies for JRFs. One was the candidate of Agnanam his caste-man. The second was a student of Prof. Lakshmanan.

During this time, I came across a student of Agnanam, who was interested in Pesticides. He was internationally connected and active as a hobby! Thus we too had a common interest, Due to our work for nearly a decade, of appeared that the impacts of pesticides in agri - fields could be more consequential than hither to felt.

One day Manavalan brought a circular of PPST of a conference by them in Thailand and asked me to contact. I wrote about our work to them. The organizer invited me to the conference with a rider that the article should precede the conference in 1985. I did "Instability in the agroecosystems due to pesticides". I went to Thailand presented it and got appreciation.

There I met 2 Malaysian Indians – Mohd Idris and Mohan. In a few months they invited me for an international conference at Penang. Here a full text must precede the conference. By this time I left Madurai to become a Professor of Environmental

Sciences at Bharathiar University at Coimbatore. I sent the paper on Pesticides with a new concept and participated in this conference at Penang for a week. This was in 1986.

Politics forced me to leave Bharathiar Uniersity and join Pondicherry University as a Professor in Ecology, In 1987. A conference was going on during early 1988; My talk on rural ecosystems has just been over. A postman came and gave me a parcel. I found there a book "Has Humankind a Future?" I found my contribution there – this is more an article than a research paper. This book was in circulation for a few weeks. That was its popularity. This article advances a postulate that the human kind suffers the ultimate consequences. These could pass through several generations, unfortunately, as gene-epidemics.

REFERENCE: Sharma, C.B.S.R., 1988. Pesticide induced gene-epidemics in the Third World countries (in) Global Development and Environment Crisis - Has Humankind a future ? (ed. Idris) :642-662. 1988.

I. INTRODUCTION

Pesticides have been in use for centuries to protect man's food supplies and help him combat the disease causing germs. Three generations of pesticides has passed by since then — Metallics, Organochlorines, Cyclodienes, Carbamates, Organophosphates, etc. using which fungi, insects, rats, nematodes, molluscs, weeds noxious to man could be eliminated. We have now about more than 400 basic pesticidal chemicals constituting up to 10,000 formulations of trade and the quest for more is never quenching! Billions of pounds of these substances are produced — all designed to be toxic to specific target groups. In every country they are always on the rise both in variety and quantum. All these were synthesized, manufactured and formulated in the Developed North and after their use or disuse, exported to the countries of the Developing South — most of which do not have enough expertise even to distinguish chaff from grain in the matter of pesticide use and safety.

Three generations of pesticides have thus criss-crossed the biosphere in quest of protecting man from pathogens and his food supplies from pests. In this process, while the goals progressively became distant, every source and sphere of human need got contaminated — a case of adding insult to injury. The ramifications of this North — generated anomaly loom large on the human society in the South also. A benumbing spectre of genetic diseases — Genepidemics — appear to be the inescapable risk the future generations in the Third World may have to face. Knowledge of these aspects may serve as a forewarning of the risks and may help in taking some mitigating steps, certain irrevocable damage having been done already. Let us examine this question in some detail.

II. DISTRIBUTION IN THE ENVIRONMENT

With nearly one pound per individual, approximately, where do these chemicals go? They remain in the habitats or crops or users or survivors among the targets and anywhere in the web of life. And they remain intact as residues or modified to more or less dangerous substances than the parentals. Thus there could originate second or even third generation chemicals in this physical and biological mileu.

The residues of pesticides and their derivatives pervade almost all the habitats known to man. All agricultural lands of the world have these substances as a rule. Most aqua cultural habitats — both marine and fresh water, and the potable water, whether surface or underground are polluted by them either directly or indirectly. Most grains, fodder, vegetable, meats, milk, dairy products, fish, eggs and almost every variety of food have traces of pesticides. So complete is their pervasion that it gave rise to the cliche — "Permissible levels". No human habitation is devoid of these substances — tropics, subtropics and temperate regions. Even the Polar Regions are not exempted from these. All tropic levels are afflicted by them, getting biomagnified in the process rendering MAN as the ultimate target of all the remainder of pesticides after the kill is over. A clear cut case of BOOMERANG. The man, for whose long-lasting safety these pesticides were designed, ultimately became their target with the pollution of his habitat, water and food.

III. TOXICITY

A. Insects

While pest populations are depleted through the application of pesticides many useful insects are also eradicated totally or reduced in density. Predators of pests are also threatened. Even the pests developed resistance. Today we have about 400 new strains of resistant insect pests and the number is ever growing.

B. Birds

As food chain terminals, birds are major victims. Apart from the population depletion, a whole species could be endangered. Egg shells often became defective. A kind of behaviour where the mother breaks its own eggs is a symptom that could be induced by pesticides.

C. Fishes

The number of episodes where shoals of fish were reported dead are a legion. Even the survivors are harmful because they could carry the pesticides to the human system. Thus, a depletion of supplies and the contamination of the available supplies resulted in several places in the Developed North. The same has been true in the South especially India.

D. Man

In man the toxicity can arise not only due to ecological exposure through food and water but also to association with pesticide contaminated environments — production, storage, distribution and usage. Toxicity to respiratory, circulatory, dermal, neural and reproductive systems can result. All the above effects are very well documented in literature during the last three decades.

E. Others

Similar toxic effects are known in every conceivable plant and animal groups. However, pests to man's resources undergo genetic changes due to pesticides developing resistance to pesticides. The importance of this knowledge is that man can similarly be affected if sufficiently exposed.

Two questions arise now. What is the enormity of pesticide exposure in the human society especially in the Third World? How is the Genetic Disease scenario in the human society?

IV. GENOTOXICITY TRENDS IN THE HUMAN SOCIETY

Though the possibility of human diseases developing from defects in the genetic material has been known for long. It is only after the chromosome technology has improved that reliable information has been forthcoming. Now it is known that cancers, birth defects and genetic diseases in man have origins, in the genetic material — genes, chromosomes or genomes. For nearly two decades scientists made estimations of genetic defects in man. It is now agreed that about 10-11% of all live births carry genetic defects detrimental to their well being and ultimately constituting a social burden. These defects have variable potential to pass on to the next generation. But the projections are that they will always increase with each generation unless something spectacular is done to prevent them. These may be chromosomal — numerical/structural, sex-linked / autosomal or genic-dominant/ recessive/multi-factorial in nature. Most of these are based on studies in the Developed North. The causes arc supposed to be chiefly — drugs, dyes, cosmetics, foods, pesticides and pollutants in the air and water. The above estimates exclude cancers which on their own are projected to afflict 25% of the population by the turn of the century

In the South very few studies have been carried out on these aspects. But the information available stresses that it will be a repeat performance here also. Genetic diseases are on the rise. Birth defects are on the increase. Cancers in every country appear to increase every year .

The intriguing question is: How could genetic effects be on the rise in South? When the chemicalization of life is far less in the South compared to the North? I attempt to answer this by proposing a thesis.

V. THESIS

Cancers, birth defects and other genetic diseases are on the increase all over the world due to a great variety of chemical agents. But in the developing South, pesticides alone are omnipresent for a variety of reasons and non reasons. These are already known to carry the highest stress index due to their all pervasive nature in the human environment leading to an acceleration of mutation load in the human populations. In the next decade or so, there could therefore be an explosion of carcinogenicity, teratogenicity and mutagenicity in the third world countries even surpassing the projections for the North. These genetic changes are irrevocable and without any warning signals. The appearance in a recognisable form would mean that the damage will have taken its fullest toll - sometimes for many generations, This thesis is based on the data from the North and ground facts on pesticide situation in the developing countries.

A. Exposures

Prior to examining the ground facts in the third world, an understanding of the possible ways of pesticide entry into man is essential.

(a) Residues: Traces of pesticides in water, food and other consumables may directly enter human system. Disinfection of contaminated water through chlorination would increase the pesticidal species both in numbers and quantity.

(b) Derivatives: Newer species of these toxins can be formed due to natural degradation or interactions with other substances or due to pyrolysis.

The metabolites of Difiubenzuron, Mirex etc. are known to be more hazardous. There could be many more such cases among the few hundreds of pesticides at large.

Pesticides are known to be capable of undergoing about 40 reactions. Also they could react with nitrogenous substances in the body. The nature of these derivatives and their hazard potential are not much known.

During the cooking of contaminated foods such as meats, fish, milk and vegetables the pesticide residues may also be modified just as the amino acids and sugars and gradually accumulate in man with time.

(c) Impurities: Contaminants in the pesticide itself are known to have Genotoxic potential.

B. *Generation of Cryptomutagens*

More ominous than the direct or indirect genotoxicity is the possibility of the generation of cryptomutagens in the human body under pesticide influence. Many pesticides are known to induce liver metabolism thereby potentiating it to activate other innocuous chemicals in the digestive system into genotoxins. Therefore if some pesticides are present in the liver even in traces, basically harmless substances could be rendered hazardous. Through this route, pesticides become instrumental in the enormous acceleration of genetic defects as currently predicted.

C. *Pest Resistance and its Relevance to the Thesis*

Another support of my thesis on Gene-epidemics comes from the well known case of pest resistance to pesticides which is genetic in origin. Mutations to this effect are caused at low dosages. Similar or less doses pass through the food chain and accumulate with time. Therefore mutations could be expected to take place all along the food chain; only they are not adequately monitored excepting in the experimental systems. In other words, if mutations can take place in the insects due to pesticides similar events could as well take place in humans too, the seat of the highest biomagnification for being the end of the chain.

D. Is the Picture too Gloomy?

It may be agreed that the picture is painted too dimly, for, most pesticides after all are degradable. Well, that little time is adequate to do the damage to the genetic material. And the expression of this damage could well take a few generations — too painfully slow to suspect the culprit. Therefore it is necessary to depend upon the Northern studies and extrapolate to support the pathetic environment of ignorance and apathy in the South. Then the message that epidemics of genetic origin are in the waiting cannot escape any concerned person, however empirical the conclusion may now appear.

E. Third World Efforts

Unfortunately, most countries in the third world have not applied their minds to these problems. Even in India where considerable expertise exists, the ground data are painfully insufficient. A whole field shrouded in ignorance, nevertheless harbouring many genetic mini - time bombs, awaits exploration and propagation.

VI. THIRD WORLD SITUATION

A. Production and Usage

Most countries in the South do not produce pesticides. They import the active principles from the North for local formulation. With the projected doubling of food output by the turn of century, pesticide consumption is also expected to increase likewise. This increase in the third world is about 9% annually now. Most of it is contributed by the Developed North as exports which now stands at about 30% of its total production. Chief contributors to this situation are India, Mexico and Argentina. A major consequential casualty would be the inland fisheries of the third world — loss of proteins due to additional of toxins, in favour of a higher carbohydrate consumption.

A large number of pesticides banned in the North have found their way to the Third World Market either deliberately or due to ignorance. Many of these introductions are unjustifiable. Undesirable impurities are also known to occur in these imports either due to defects in manufacture or over-storage. A growing trend is the launching of enterprises in the third world by the Principals and with easy morals.

The tendency of the recipient societies to overuse these toxic substances in their rush to protect their meagre food supplies coupled by a matchless ignorance, work as a spur for the pesticide industry to make a quick buck.

B. Misuses

The pesticides are often misused — to protect the stored grains and fruits where they were not intended. Several fruits and large vegetables are given a "Dip of Insurance" in the pesticide solutions. Pesticide drums are often reused for storing foods such as oils. Jaggery production in some quarters has a step to bleach it with the use of some pesticides! One may absolve the industry by arguing that all these were done against the instructions. When the people are ignorant, why cannot their sales and applications be controlled and monitored by appropriately trained agencies? This would naturally upset the vested interests — a formidable combination of Politicians, bureaucracy, Industry and Tradesmen.

C. Exposures

In man's eagerness to protect his food supplies accompanied by lack of guidance a myriad problems are generated — residues, food chain contamination and magnification, pollution of surface and underground waters and finally the more serious occupational exposures which may happen during formulation, storage, marketing and application.

D. Episodes

Countless accidents of varying dimensions occurred all over the world during the last 30 years. In India the case of Bhopal is quite fresh, for, while the fatalities are over, long lasting and heritable effects are slowly coming to surface. Handigodu (in Karnataka, India) syndrome also has similar lessons unfortunately not heeded. Recently in Andhra Pradesh (India), farmers committed suicide due to economic unbearability of crop losses in cotton. This epidemic loss has been traced to the development of resistance by white fly and American hopper to synthetic pyrethroids among other causes — another case of genetic change. Vietnam has offered a classic example on its own. The lesson is that pesticides at lower doses can expose the biota in a chronic fashion resulting in a very sinister and unapparent effects — the genetic effects

E. Apathy and the Unholy Alliance

The apathy in the Third World towards pesticide hazards is disquieting and is due to the cumulative effect of a variety of reasons — the most common being ignorance. But hiding these bitter truths by the North and their counterparts in the South-Politicians, Industry and Bureaucrats is astonishing.

VII. FINAL DIAGNOSIS

In the final analysis pesticides are not acceptable. One must also realise however that there will be serious consequences if they arc totally banned. We have to prepare for these consequences. Immediately, as a measure of mustering strength to declare a war on this mischievous menace, several steps can be taken.

A. Research

As stated earlier, there is hardly any work in this field in the third world. Unimpeachable evidences are still needed to unequivocally implicate pesticides in the spread of genetic diseases to the extent feared. For this, extensive surveys and intensive researches on clinical data, hospital records and population surveillance have to be taken up in all the countries of the third world. Then alone a clear understanding of the incidence of the birth defects, genetic diseases and cancers in these parts can develop. Concurrently, state of pesticide situation in these countries has also to be monitored though the available evidences are quite ominous.

B. Education

Education of Masses about these "irrevocable hazards without warnings" by Governmental agencies is necessary. Inclusion of this branch in all courses of Medicine, Agriculture and Environmental Science might prepare the agents of change to cope with this eventuality. Awareness campaigns and alerts by the Non-Governmental agencies and media would arouse concern among the people.

C. Controls

Replacing synthetic pesticides with botanicals will ease the situation to some extent. Phasing out pesticide use and ultimately banning them could be a very ideal goal. But given the realities such an eventuality would be a mere dream. At least in the third world they can be banned and the crop "losses made up with imports from North — A Polluter — Pay - Principle!. In the interim, no pesticides should be freely available. And they should be applied only by the trained personnel of the companies and governments or the authorised marketers.

D. Interactions

Global network to collect and collate information on all the aspects detailed in this paper could go a long way in alerting the global conscience and influencing the international bodies. Political will is the ultimate need. Either the environmentalists should enter the public life or the political parties have to be environmentalised. This is a matter for discussion.

VIII. FOREBODINGS

One must remember that Gene Epidemic portents would mean

(a) at least 10% of gene abnormalities in the human population which will increase with each generation and

(b) 10-25% of cancer subjugation and ultimate mortality through unprecedented suffering.

Some estimates reach up to 60%, both put together! All these constitute an enormous social burden on the already fragile economic fabric of the third world. A revolutionary change for the worse in the health condition of mankind in the third world is clearly predicted. In the least, the issue is highly emotive. With populations on the rise, food supplies faltering, the spectre is indeed frightening and a shock or a disaster in this direction is on the cards. Pestilence stealth in darkness!

If Malthus and Darwin's postulates on population were averted by modern science, the products of the same science has portents to unleash epidemics of genetic origin and prove these prophets tellingly true unless intervened by a nuclear holocaust in between.

As much data as possible on pesticides has been appended, such as: Pest resistances, Genotoxicity, Carcinogenicity, Mutagenic Structures, generation of Mutagens, Genotoxic Impurities, ransformations, Pesticides banned in the First and Third Worlds.

10

INDUCED ANEUGENESIS IN PLANTS

While working on the review on the genetic effects of Pesticides some data generated by our students did not tally with others and appeared strange. So I kept them aside for the time being. after the approval of my pesticide review by the CRC Press, On re-looking, a pattern emerged - a pattern of aneuploidy in plant systems under the influence of pesticides. In other words some pesticides affect only the spindle mechanism that segregates the chromosomes into different cells. While aneuploidy could be useful in plant genetics, it is very harmful in human systems. Therefore a full fledged review was generated without ever planning.

First I communicated it to Dr. Irene Adler at Munchen, Germany, for publication in the Journal she edits – Mutagenesis, Carcinogenesis and Teratogenesis. It was approved, but with an information that one of the referees from England liked it in his Journal, MUTAGENESIS. I readily agreed. This editor Prof. Parry send it to a new set of referees for criticism. Reply came - nearly 5 referees approved it with hardly any changes and published.

At Cleveland during the 3rd International Conference on Environmental Mutagens, I was intently listening to Dellarco's lecture on aneuploidy. Then I found a pat on my back saying "He has not covered the points you raised" very appreciatively. He was Prof. J M Parry of Swansea University and editor of MUTAGENESIS. On another day I chanced upon crossing Dr. Irene Adler coming in front of me. She stared at my chest to know my name and exclaimed " you are Sharma ! who rote the article on Aneugenesis. I never expected that such an article from India. Then came Eliezer Hubermann, shook hands fervently after seeing my name and told Irene that I lost a chance to work with him at ORNL, due to governmental policies". She was a German and he a Jew from Israel !

REFERENCE: Sharma, C.B.S.R., Chemically induced Aneuploidy in higher plants. Mutagenesis 5:105-25. 1990, UK.

Although there is extensive literature reporting aneuploidy in plants there has been little consideration of the mechanisms of induction of the observed changes in chromosome number and the role of environmental agents in its induction. This paper collates and analyses the literature on aneuploidy induction in plants with the aim of identifying the mechanisms involved. Many of the major events that occur during cellular segregation, such as respiration, spindle and phragmoplast functions and chromosome doubling, appear to be stressed by agricultural chemicals, drugs, natural and industrial products, leading to aneuploidy in both the mitotic and meiotic cells of 60 plant species. Of the nearly 400 chemicals at present suspected to be aneugenic to plants, 150 are agricultural chemicals, 60 are drugs, 60 are natural substances and 50 — 60 are industrial products. Nearly half of these agents were reported to affect spindle mechanism in mitosis and meiosis. About 60 affected phragmoplast function in root meristems. Another 60 chemicals appear to modify chiasma frequencies. Nearly 40 chemicals appear to be clastogenic in such a way that could lead to aneuploidy by a secondary mechanism. About 45 plant species exhibit aneuploidy without any apparent cause or source of stress. This could arise from several metabolic and anthropogenic causes operating alone or together such as: (i) the nutritional status of the soil, (ii) irrigation by polluted water, (iii) the ageing of seeds, (iv) pathogenesis of plants, (v) cryptotoxins in plants, (vi) pesticide applications and (vii) pesticide residues. Environmental factors such as pH and temperature may further compound these stresses. Among the biota, plant systems alone seem to possess all the components to monitor every conceivable type of aneuploidy and respond to all kinds of stress in our environment. There are deficiencies in the work hitherto carried out which need to be rectified. Some aspects have to be reinvestigated with improved protocols. For a perspective understanding of the phenomenon of aneuploidy, so vital to human systems, it is imperative that we devise new approaches, some of which are outlined here.

INTRODUCTION

The detection of aneuploidy in plant species of economic importance received considerable stimulus with the observation of the induction of polyploidy by chemicals. The need to screen potentially harmful chemicals in the environment led to the screening of chemicals in various biological systems for their genetic effects. Plant systems, amongst others, were also used as screening assays. Thus, the early objective of inducing genetic changes in plants for agricultural exploitation led to a quest for methods to screen out such chemicals. These twin objectives have generated a considerable database on the responses of plant cells to chemicals.

The division of cells in higher plants is limited to the meristematic and sporogenous tissues and consists of karyokinesis and cytokinesis preceded by G_1, S and G_2 phases.

During G_1, S and G_2 phases the synthesis of DNA, RNA and proteins necessary for the division of a cell is carried out. Following this, the coiling of chromosomal fibres occurs, leading to a gradual shortening of chromosomes forming chromatids. These are subsequently separated into daughter chromosomes which move to opposite polar regions of the cell. A division of the cytoplasm is the final act in these continuous processes, ultimately leading to the formation of two independent cells. During these processes any interference at any stage may lead to abnormal appearance and/or behaviour of chromosomes which may be apparent immediately or in the succeeding generations. This interference by chemicals and radiations in the environment has been the subject matter of genetic toxicology and environmental mutagenesis.

While considerable literature exists on the biochemical, mutational and clastogenic aspects of chromosomes, the numerical chromosomal endpoints have been comparatively neglected. Therefore, information on effects indicating aneuploidy are scattered throughout numerous publications, many of which deal with spontaneous occurrences. This present paper aims to gather together these scattered reports to provide a coherent review of this important subject.

Although aneuploidy was reported a considerable time ago in Avena, Oenothera and Datura species, it was only after the concerted efforts of Blakeslee that this aspect received attention. The first major review of this topic was produced by Khush.

Definitions

The term 'aneuploidy' was first introduced by Tackholm

and a number of other terms such as primary, secondary and tertiary trisomes, sectorial aneuploidy, monosomy and nullisomy, trisomigenesis, etc. are in use. However, it is necessary to introduce alternatives more in tune with the existing terminology in genetic toxicology. Thus the term 'aneugenicity' is adopted here to denote the generation of aneuploidy, with 'aneugens' for the generating agents. Aneugenesis produced through a single-step mechanism may be considered direct and when it occurs through two or more steps, indirect. The word 'trisomy' in the terms primary, secondary and tertiary trisomies, in vogue in plant genetics can, however, be retained without change. All this appears semantically in tune with terms such as mutagenicity, clastogenicity, clastogens, mutagens, etc.

Information on the aneugenicity, assay systems and endpoint contained in this review was obtained from primary and secondary sources of information. Any publication where aneuploidy is reported has been considered. Repetitions are avoided unless otherwise necessary and negative results were excluded from consideration.

MECHANISMS OF ANEUGENESIS

Chemicals appear to act on a variety of sites in dividing cells to cause an addition or a deletion of a chromosome, in full or in part, either directly or indirectly. Many of these mechanisms are common to both mitocytes and meiocytes while a few are special to the latter. All these mechanisms can be broadly grouped into four types.

Respiratory inhibition (pseudo-aneugenicity)

Some substances temporarily asphyxiate cellular respiration and paralyse spindle function. This results in the arrest of cells at metaphase, which, however, disappears following the removal of the toxicant from the cellular environment. Such substances are large in number and variety, and some are incorporated as pretreating agents in karyometric studies. These temporary toxins, often erroneously referred to as spindle poisons, are merely pseudo-aneugens. It is likely that most reports of metaphase arrest, especially from pesticides, could only be from this route.

Spindle dysfunction (direct aneugenicity)

The roles of tubulin, microtubules and spindle mechanism in the anaphase segregation of chromosome in the dividing cells are well understood and documented and do not merit repetition. However, a brief mention may be appropriate. The equilibrium between tubulin dimers and the already formed microtubules may be affected as in the case of taxol; an already formed spindle may be split causing multipolarity as in the case of griseofulvin; the microtubules may be disassembled by binding at any of the three sites leading to vagrancy as in the case of colchicine and vinblastine; bivalents may not congregate at the metaphase plate as in the case of some pesticides.

Substances affecting any of the aforementioned processes are known to elicit immediate cellular responses that are generally expected to lead to subsequent ploidy variations. Quite often, at concentrations above the threshold level, many toxins — especially pesticides —may physically disintegrate the spindle, thereby arresting the metaphase.

Phragmoplast dysfunction (indirect aneugenicity)

If the chromosomes have accurately segregated into the two halves during the preceding stages, dysfunction of cytokinesis through an interference of phragmoplast formation will not directly lead to aneuploidy, but will lead to binucleation. When these two nuclei begin to complete the next cell cycle within the same cell (i.e. bimitosis) the anaphases may mix up due to a sort of 'crowding effect' from spatial limitations, resulting in nuclei with abnormal numbers of chromosomes. Unfortunately, no attempt has so far been made to investigate this phenomenon. Therefore the possibility of aneuploidy through this indirect route cannot be ruled out as there are more than a score of chemicals known to affect phragmoplast and cytokinesis in mitotic and meiotic tissues.

Chromosome 'dysfunction' (direct and indirect aneugenicity)

The chromosome is the site of three kinds of events when exposed to chemical stress which may lead either directly or indirectly to primary, secondary and tertiary aneuploidy.

The first pertains to the longitudinal separation of the chromosome, which may directly lead to primary aneuploidy.

(i) If the synchrony of division and disjunction of the centromeres is lost in any chromosome, the simultaneous segregation of chromosomes will be disturbed leading to vagrancy, ultimately resulting in primary aneugenicity, i.e. trisomy in the daughter cells. This source of error has received attention.

(ii) Another way for such a change is the extra replication of one or a few chromosomes which naturally leads to an extra chromosome. Even in meiosis this could happen in a half bivalent in response to high temperature and phytohormones.

(iii) Similarly the centromere of a half bivalent may prematurely divide while till constituting the bivalent. During meiosis I segregation, this leads to a monosomy in one resultant cell and trisomy in the other, much as happens in response to ethylene glycol in maize.

The second kind of event pertains to clastogenicity in the arms and centromere of the chromosomes. This may indirectly lead to primary, secondary and/or tertiary aneuploidy.

(i) If sub-chromatid exchanges occur in mitotic metaphase, the chromosomes may not dysjunct even if centromere division is completed, leading to aneuploidy. This could similarly happen during metaphase II of meiotic cells. Such incipient clastogeny has not received much attention though stickiness, which could be a manifestation of this phenomenon, has been widely reported.

(ii) On the other hand, a transverse division of the centromere, misdivision, which is often seen as an 'extended centro-mere', will result in the formation of isochromosomes. Similar breakage and exchange of chromosome segments leads to the origin of translocated or repatterned chromosomes. These aspects are too well documented in the literature to warrant further repetition.

The third kind of event concerns the chromosomal pairing and chiasmata functioning *vis-à-vis* spindle activity in the meiotic cells. These may directly or indirectly lead to primary aneuploidy.

Chromosome conjunction, bivalent congression and dysjunction mechanisms could be affected leading to asynapsis, non-congregation, desynapsis, early dysjunction, delayed dysjunction, or non-dysjunction of bivalents. In other words, a non-conjunction of chromosomes or a faster terminalization of chiasmata results in a precocious release of half bivalents. Conversely, slow terminalization of chiasmata and non-dysjunction of bivalents will result in a delayed release or a total non-release of half bivalents. Further movement of these elements could be abnormal leading to laggards, forwards, micronuclei and aneuploidy either immediately or in the subsequent division. Such events have been reported to happen in response to temperature variations, col-chicine and some pesticides.

CELLULAR INDICATIONS

Aneugenesis is indicated in dividing cells in many forms. Many indicators are common to both mitotic and meiotic cells while a few are specific to the latter. All these are caused by one or more of the mechanisms described earlier either directly or indirectly. These are briefly described.

Mitosis

When the segregation mechanism of anaphase is affected, the consequences are seen immediately. All chromosomes may be arrested at the equatorial plate (i.e. metaphase arrest).

On the other hand, some anaphase chromosomes may be scattered (i.e. scatters), one or a few of them may reach the poles precociously (i.e. forwards) or lag behind (i.e. laggards) while the remainder are segregated to poles in a synchronous manner. Sometimes the two daughter chromosomes may remain joined due to non-disjunction of the centromere without moving to either pole (i.e. diplochromosome). These together may be termed as 'vagrancy'. Occasionally, the anaphase groupings increase from the customary two as a result of disturbed polarity (i.e. split-spindle). Except for the arrested metaphases, all others may form micronuclei. However, when the phragmoplast is affected, cells become binucleate and during the ensuing bimitosis there could be an unusual mix up of anaphase chromosomes due to spatial limitations (i.e. crowding effect). Therefore binucleation and bimitosis are also suggestive of a likely aneugenic consequence.

Thus, scatters, forwards, laggards, diplochromosomes, multipolar groups, micronuclei, binucleation and bimitosis indicate primary aneugenicity in mitotic cells. Secondary aneugenicity is suggested by centromere misdivision and probably by an extended centromere, as it could be a prelude to misdivision. Micronuclei also indicate secondary or tertiary aneugenicity if they are organized by chromosome fragments rather than whole chromosomes.

Meiosis

Meiocytic responses to aneugeny are similar to the mitotic responses. A bivalent may not align at the metaphase (non-congression). During anaphase it may not disjoin into half bivalents (non-dysjunction) or do so only slowly (delayed dysjunction). Sometimes due to non-conjunction, fast terminalizations or low chiasma frequency or desynapsis, univalents result which could behave as laggards or forwards. All these could also lead to micronuclei if not reconstituted to normal nuclei. Multivalent chromosome associations, ring chromosomes and isobivalents indicate that secondary and tertiary trisomics could subsequently arise.

Thus, non-congression, non-disjunction, delayed disjunction, univalency, laggards, forwards and micronuclei are the indicators of primary aneugeny in meiotic cells. The secondary and tertiary aneugeny is suggested by multivalency.

The direct and indirect aneugenic indicators in the mitotic and meiotic cells detailed above may ultimately lead to primary, secondary and tertiary aneuploidy. These manifestations could arise in response to known chemicals or due to yet undefinable chemical stresses and factors which will be considered next.

ANEUGENIC CHEMICALS

A very large number of chemical substances have been found to cause aneuploidy in the cells of 60 plant species. A few of these are also reported to cause centric, segmental and incipient clastogeny which could lead to secondary and tertiary aneuploidy. Barthelmess estimated that there were- 20 such aneugens. However, Sandhu identified only — 20 from publications considered by them to be reliable. Yet a realistic figure may be much higher than Barthelmess's. These may be broadly grouped into four categories: (i) agricultural chemicals, (ii) drugs, (iii) natural products and (iv) industrials and others. A brief summary should be adequate.

The effects of a few of these chemicals stand out for the thoroughness of investigation and for their frequency of occurrence in literature. These include colchicine, vinblastine and benomyl, ethylene glycol, caffeine, griseofulvin, formaldehyde and ammonia, lindane, nitrous oxide, halothane and methoxyflurane, myleran and ethoxy caffeine, chloramphenicol, taxol, tetracycline, chloral hydrate, ethidium bromide, ethyl methane sulfonate, cyclophosphamide, p-fluorophenyl alanine, and diethyl pyrocarbonate

Agricultural chemicals

Between 140 and 150 pesticides and growth regulators appear to have aneugenic potential in 40 — 45 plant systems. The anaphase segregation mechanism in root meristematic cells of 40 plants, resulting in vagrancy, was affected by 130 substances; 17 of these caused similar effects in the embryonic shoot cells. It is, however, not known whether the impact is on the spindle activity or centromere division or both. Over 40 chemicals showed a potential to impair cytokinesis in 10 species that could lead to binucleation. However, only very few of these are known to lead to bimitosis. About 70 chemicals appear to affect the meiotic spindle in > 17 plant species. Of these, nearly 40 are reputed to also affect the chiasma mechanism in seven plants. Furthermore, 30 chemicals are reported to cause incipient clastogeny in eight species that could secondarily lead to aneuploidy, though no firm evidence exists at present.

Drugs

Between 50 and 60 drugs appear to have aneugenic effects in 10 -15 plant systems. Most of these act predominantly on anaphase segregation in the root meristem cells of seven species. Only a few of these affected phragmoplast functioning (Table VIII). In meiosis, spindle and chiasma mechanisms were affected by four drugs. Clastogenicity that could lead to secondary aneugenicity has also been recorded.

Natural products

Between 50 and 60 defined chemical substances of plant origin induce aneugenesis in 15-20 species, in addition to several undefined extracts of seeds, leaves, fruits and stems. Fifty of these chemicals caused spindle dysfunction in > 10 species. Cytokinesis was affected by nearly 10 of these. In meiosis, spindle and chiasma mechanisms were affected only by a few of these chemicals. Clastogenicity that could lead to secondary aneugenicity was also caused by some of these substances.

Industrials and others

Man-made chemicals other than agricultural chemicals and drugs are grouped here. Initial components or raw materials, intermediates, end-products, byeproducts and waste products of various industries have not been widely screened in plant systems for aneugenic activity. Only 50 -60 chemicals appear to have been screened in 25-30 plant systems. Nearly 50 of these caused spindle dysfunction in 15 species. Cytokinesis in about five plants was affected by nearly 10 chemicals. In meiosis, the spindle mechanism appears affected by > 15 chemicals in < 10 plant species. Clastogenicity that could lead to secondary aneugenicity has also been recorded.

In the case of spindle dysfunction, whether or not the chemical effect is on spindle activity or on centromere division or both could not be delineated for any agent. Similarly, whether metaphase arrest was due to respiratory asphyxiation or a genuine spindle dysfunction was rarely considered. Also, the consequences of binucleation in the succeeding cell generations such as bimitosis and anaphase crowding were rarely followed up

Thus, of the nearly 300-350 chemicals presently considered to have an aneugenic potential, not more than 20 were studied in detail

It is necessary to realize that these agents, barring a few like colchicine, were not tested exclusively for aneugenic effects. These effects were noticed along with clastogenic effects. Further, this aneugenic activity was rarely quantified, except in a handful of publications. They are usually grouped with clastogeny and labelled as aberrant cells. Consequently, in the information given in the text and the tables some omissions cannot be ruled out.

ANEUGENIC FACTORS

Aneuploidy has been observed to occur spontaneously in many crop plants. This suggests that operation of some hitherto obscure chemical influences on the cellular segregational mechanisms which may be natural and/or anthropogenic. All these could be further influenced by environmental factors.

Nutritional status of the soil

The nutritional status of the soil may have some impact

on cell division both in the root and in the shoot systems. The effects on the shoot systems may be accentuated as development progresses and may manifest themselves in the germ line. Salts such as phosphates and nitrates, metal ions and organic matter may be implicated. This has been shown in the case of chiasma frequency and other end-points in Tradescantia and Hordeum.

pH, which may vary with soil type and constitution, may also influence the effects of chemicals on cells.

Irrigation by polluted water

Irrigating agricultural fields with polluted water, especially domestic sewage, with or without treatment, subjects agricultural ecosystems to chemical stresses. Reports of genotoxicity of such wastes are becoming more frequent. Aneuploidy could be among such genetic effects. Naturally occurring plants and animals are also known to be genetically affected by polluted environments. Even vegetables grown with sewage are suspected to harbour genotoxins. Given the large numbers and ubiquity of these toxins from waste-waters, it would not be surprising if many of these act as aneugens in the plants irrigated by waste-waters.

Pesticide applications

The application of insecticides, fungicides, herbicides, growth regulators and chemical fertilizers on crops are a part of modern agriculture. Some of these chemicals were designed to be poisonous to pests and pathogens. When these come into contact with the growth and/or reproductive parts of plants, cell division is stressed.

The practice of treating seeds with fungicides to prevent infection during storage might lead to aneugenesis in embryonic tissues. While the effects may appear in the seedlings as temporary toxic symptoms, they could also give rise to aneuploidy in the adult plant. Similar events may take place due to the spraying of pesticides before, during or after the emergence of the inflorescences. Experiments with individual agents have demonstrated such a possibility. However, in field conditions more than one chemical is used, which may multiply these effects.

Pesticide residues

A variety of pesticide residues reach agricultural soils year after year in varying amounts, and these may be converted to other chemical species. These are likely to be taken up by the plants and cause variations in the behaviour and the numerical status of chromosomes. Some evidence to this effect is available from the cytogenetic behaviour of weeds in crop fields with a pesticide history. In fact a theory that pesticides may promote evolution by inducing genetic variations in plant populations was proposed some time ago. That genetic effects in the biota of the agro-ecosystem have the potential to bring about instability in the whole system has been suggested. Pesticidal residues in soil can cause aneuploidy and other effects in weed and crop plants. Only circumstantial evidence favours this possibility, because no work has been done on the genotoxic impacts of pesticides on plants at doses similar to the residual levels in crop lands.

Cryptotoxins in the plants

It is known that pesticides taken up by plants remain as

residues or may be metabolically modified. Although the nature of the derived products is not known for most pesticides, their hyper-activity has been demonstrated in some plants by microbial coupling assays for mutagenicity. Therefore these could also cause aneuploidy as well in the host plant. In other words, since the origin of such cryptomutagens through plant metabolism has been demonstrated, the possibility that some of these act as aneugens cannot be ruled out.

Pathogenesis of plants

Plants are often infected by nematodes, insects, fungi, bacteria, viruses, etc., resulting in a loss of productivity. Toxins from some pathogens are known to be genotoxic in experimental systems and could therefore cause aneuploidy in the host plants also. These aneugenic symptoms may be carried by seeds of such plants. The available information relevant to this aspect from viral, bacterial, fungal and nematode infections and toxin impacts on plant cells lends strong evidence in support of this possibility. If the seed outputs from the infected populations are screened for chromosomal variations, pathogenic induction of aneuploidy in the plants through endogenous toxins may be observed.

Ageing of seeds

Ageing of seeds during storage can alter the metabolic processes and accumulate antimetabolites within the seed. During germination these antimetabolites can cause aneuploidy in the root and shoot primordia, among other effects, which may be transmitted to the adult plant. Aberrant chromosome behaviour that could lead to aneuploidy is known to occur in many plants through this route, though some of these appear to be eliminated in the germline in a few plant systems.

Environmental factors

Humidity, light intensity and ambient temperature are known to alter the flowering process. Therefore it is logical to suspect that meiotic events are also modified to some extent. While gross cytological changes may not happen, prophase errors such as asynapsis, desynapsis and altered chaisma terminalization could result. Therefore some spontaneous variations in numerical stability may occur through this route 2). Furthermore, any of the aforementioned channels of indirect aneugenesis can be influenced by these environmental factors.

Most of the spontaneous aneuploids detected in as many as 65 plants since 1915 may have in fact arisen through one or more of the above mechanisms. The chemical agents actually responsible have not been identified since aneuploidy research itself is still in a formative stage.

Perspectives

Genotoxicity research in plant systems has resulted in a considerable literature. But it has not received due recognition owing to a few short-comings, though a great deal of concordance was established between plant and mammalian responses to genotoxins). It may be therefore necessary to reorient the existing studies pertaining to test systems, use of test agents, principles of protocols and other aspects with special reference to aneuploidy problems.

An understanding of the short-comings in the current works might help develop better practices. In this process, a few new approaches could also emerge for consideration. These are briefly examined in the following.

Aspects of concern

The information available on aneuploidy in plant systems is defective in a number of ways.

(i) The aneugenic endpoints in the cells were not well defined and delineated from clastogenicity. Consequently they were seldom quantified.

(ii) As a result, the dose-responses of these aneugenic endpoints, either collectively or individually, were not measured.

(iii) Although most studies combined aneugenicity and clastogenicity as 'aberrations', whether or not thresholds occur between these two kinds of endpoints vis-a-vis the administered doses of causative agents was not examined.

(iv) The primary distinction between aneugens and pseudoaneugens has rarely been attempted, though it is known that cellular respiratory asphyxiation can cause temporary and reversible metaphase arrest.

(v) Despite their developmental relevance, shoot meristems were rarely chosen as the major subject of testing.

(vi) Meiosis was generally not investigated in depth for aneugenic effects.

(vii) Last, but not least, preparation of test solutions, choice of target tissue and its growth and maintenance, its treatment and recovery and the time periods involved in all the above have been defective to different extents.

All these can be rectified by modifying and rationalizing the existing practices.

Agents and doses. The basic properties and background toxicity data of the test chemicals have to be taken into account.

Temperature and pH should also be controlled. The dose ranges need to be determined by conducting preliminary experiments on cytotoxicity responses of the dividing cell systems and proportionately diluting the toxic dose. The time and duration of treatment and recovery period should depend on the duration of the cell cycle. Thus the test dosages should range between the cytotoxic threshold and the minimum effective dose. Treatment of the target organ should be of minimum duration based on permeability data and the recovery period should be proportionate to the duration of cell cycle). Recovery after treatment should always be in a nutrient solution. All solutions should be aerated whenever the test system is immersed in them.

Systems and assay protocol. The plant systems that can be used are well known with only a few likely additions, e.g. Arabidopsis, Allium, Pisum, Ephedra, Vicia, Tradescantia, Crepis, Plantago, Bellavalia, Nigella, Glycine, Hordeum, Phlox, Pterotheca, Haplopappus, Eichornia, Lemna, etc. The organs selected for exposure to the chemicals used as test agents, positive controls, negative controls and solvent controls, may be seeds, seedlings, roots, developing regions such as inflorescences and branches. The tissues of study need not be only the primary roots but also the secondary root meristems, embryonic shoots, pollen and embryo sac mother cells, pollen endosperm and embryos.

The exposure of target organs - which may be periodic, acute or chronic-can be by immersion, injection, spraying and incorporation into culture medium. The younger the plant system, the greater will be the opportunity for diplontic selection of the impact of the treatment. This approach will minimize effects at meiosis and gives a realistic picture of ultimate effects. Thus, following up the line of transmission until the end of a generation is very desirable. Alternatively the assay should include a long transmission across one or two cell cycles (i.e. with appropriate recovery periods) or until the resultant organ is formed (i.e. monitoring in the secondary root meristems following exposure of primary roots).

Mitotic endpoints. Aneugenic endpoints in mitotic cells have to be grouped separately from the clastogenic ones. Since both types of effects could occur simultaneously, the basic mechanisms of origin should attract major attention followed by a search for a delineation of thresholds between clastogenic and aneugenic endpoints, *vis-a-vis* the incident dosage, which is a combination of the concentration of test solution and the duration of treatment.

Meiotic endpoints. In meiosis, endpoints have to be segregated into two groups within a single study following subjection to chemical stresses. The first group consists of non-conjunction, non-congregation, non-disjunction, early or delayed disjunction, chiasma reduction and univalents, as they all lead to primary aneugenesis. The second group consists of clastogenic endpoints such as the formation of isochromosomes and multivalents, as they lead to secondary and tertiary aneugenesis. The chemical treatment may be given either during the first cell division in life history at the seed level or after the initiation of the inflorescence. In the latter case the methods of chemical admin-istration and recovery have to be carefully standardized, for they could differ from species to species.

Controls. Negative controls with solvents are necessary. Sometimes this may not be a mere ritual, for some solvents — such as ethanol, acetone and DMSO —are known to be aneugenic. The problem of positive controls, however, is more complicated. With several cellular aneugenic mechanisms known there is a necessity for several positive controls. Therefore the most urgent need now is to standardize the performance of a test system against a common set of test agents with different mechanisms of action.

New approaches

The problem of aneuploidy induction will not be fully understood and appreciated unless some unconventional approaches are devised to improve current practices. A few of these are briefly mentioned here.

Centromere asynchrony. *Though the role of the centromere in chromosome segregation has long been established, investigations have rarely been addressed to non-disjunction or asynchrony in centromere disjunction in the diplochromosomes of meta-anaphase. This can be accomplished by speeding up the existing protocol since the processes involved are fast and transient. A slight excision of the tip of the root meristem will facilitate the fastest entry of the test chemical into the system.*

This should be followed by metaphase arrest by a respiratory inhibitor so as to ensure that no other aneugenic mechanism is stressed. If this can be done in systems with a small number of chromosomes the results would be clear cut.

Transmission through development. Root meristems, on which most of the work has so far been carried out, are not develop-mentally consequential to the ultimate objective of aneuploidy at the organism level. For this purpose, shoot tips are more relevant. However, developmental transmission can also be understood and monitored even in the root systems if the primary roots are subjected to the chemical treatments and secondary meristems are screened for the effects. This can be accomplished by suitably modifying the existing protocols in plants such as *Pisum sativum* and *Vicia faba.*

A suitable protocol for studying the transmission of aneugenic effects would be to treat seeds of the test system during their first division and periodically to monitor impacts in root and shoot cells and finally in the pollen mother cells. The likely consequence of a reduction or enhancement of initial aneugeny through developmental processes would be answered only by such progeny testing. This approach might be found to be relevant to human situations.

Bimitosis and 'crowding' effect. Though binucleate cells are often reported, attempts to follow them up through bimitosis and monitor the likely 'crowding effect' due to spatial congestion have not been made. This can be readily accomplished by increasing recovery durations, in the assay protocols of any of the known systems, to cover more than two cell cycles.

Micronuclei and vagrancy coupling. Although the micronuclei are generally considered as indicators of clastogenicity, they could also originate from vagrant chromosomes. While cells with clastogenic and aneugenic manifestations in meta- and anaphases have also shown micronuclei subsequently, so far no such coupling of vagrancy and micronuclei has been attempted in any plant test system. The most suitable systems for this kind of assay in meiosis are Tradescantia, Allium, Hordeum and Vicia where scoring for micronuclei in tetrads can follow scoring for aneugeny in the preceding cellular stages. The same can also be done in mitosis where micronuclei in the primary root could lead to aneuploidy in secondary roots.

Female meiosis. All the work on the meiotic system has so far been carried out in the male get ratline. Female meiosis was rarely monitored for obvious reasons. However, stress on this is often suggested by decreased seed set even when male meiosis remained unaffected. It is also logical to believe that whenever microsporogenesis, i.e. male meiosis, is affected, megasporogenesis could not have remained fully conserved. Therefore, whenever the seed output per inflorescence is affected, one may suspect that the mature seeds may carry some genetic deviation, aneuploidy being one possibility. The direct study of meiosis in the female germline is, however, possible only in a few systems such as Allium.

In situ approach. Considering the indirect sources of aneuploidy discussed earlier, it is possible that most of the naturally occurring aneuploidy could remain undetected. The underlying fact here is that there may be many chemicals in the ambient environment in the form of complex mixtures that could subject biota and man to aneugenicity. Unfortunately these cannot be understood or monitored unless an *in situ* approach to this problem is developed. Therefore crops and weeds grown in stressed habitats need a thorough investigation for genetic effects. Some plants such as Eichornia, Lemna, Typha, Parthenium and algae such as Chara and Nitella appear very suitable for this purpose. Even standardized systems can be maintained in the polluted habitats where the effects can be monitored periodically. All these are expected to reveal significant information on indirect aneugenesis

11

MUTAGENESIS IN NATURAL ECOSYSTEMS

This review was not published. But the only one planned by me as a natural corollary to the past experiences My good fortune is that this topic was not reviewed so far although data are in heaps, and myriad of papers were published.

This could not be published only because of insincerity of students and general environment in the department that prevents any serious work and abhors science! The student Anil belongs to the batch of 1988-90. The final semester is spent in a project work. He was working with me in the field of what they fancied, Ecotoxicology. Searching for an appropriate site was a major task. We found one. Readings being taken. All of a sudden the site was destroyed! The semester ends. Search begins again. We got another. The habitat is a bit marshy. So it could not be tampered with. It took another semester. During this time he became genuinely interested in this field of genetic toxicology in the field.

This dissertation was highly appreciated by all but one chap, a lecturer who commented that "it took one year". *We knew by now that the first site was destroyed by his minions that delayed our work. This gentleman got a job first and a Ph. D later on, both engineered, a practitioner of Xenophobia of the most negative kind. The student was not of his liking.*

The student was offended due to this comment, after giving him a 'C' grade in all papers of this gang of Four, out of prejudice. Then I suggested further work on literature and develop insights and visions on future developments.– especially methodologies because his work was in situ, *out doors with its attending difficulties.*

Thus all the data work are his labours. The theory and concepts as well as figures are mine. The script is also mine. All along we were greatly disturbed by colleagues and authorities. In disgust he left for greener grasses and acceptable venues. I retired from work except teaching as my enemies desired, lead by the then VC. The file went into cold storage.

This was so between 1991- 93. Some 20 years later, 10 years from my retirement, reports came of piracies of our work by the two gangs of my former colleagues. – done or doing, a kind of Necrophilism. This prompted me to take out moth- eaten files and retrieve as much as possible.

One such effort is this one. Luckily the data and concepts remain relevant to this day proving that our thoughts and actions in those times were right. And this has another evidence to our general theme that genetic toxicology of 1970s becomes Environmental Mutagenesis of 1990s with ominous portents to human society. It also justifies that there is toxicology in our daily life, not as Sigmund Freud prophesied, but much more and consequential for generations.

REFERENCE: Anil, P.C & Sharma, C.B.S.R. 1993. Mutagens in the Ecosystems –Monitoring by Plants and Animals, p. 71.

A Private Publication

A. INTRODUCTION

1. Variations

The chief characteristic of all organisms is variation. Variations may occur suddenly or gradually or both ways. They may appear as spontaneously arising without apparent causative forces. If these variations are transmitted to the next generation they may get settled in the biota. And thus biota evolves in time.

Some however will not be transmitted ending with the life of the organism. During the living span, however they may affect the harmony of the organism within and without. These variations, morphological, behaviouristic or ecological, are channeled by biochemical mechanisms. At cellular level these originate at chromosomal, genetic and/or genomic levels. All may have some deterministic forces though not intelligible to us as they concern the past. That the "spontaneous variations" found in literature are not so spontaneous is now realized. Mutagens causing them must infarct be endogenous and present in all organisms to release local variations without which evolution of biota could not have taken place. It is increasingly felt however that the forces in the present times are mostly exogenous. This knowledge is now text book material referring to such illustrious men of biology such as Darwin, Mendel, De Vries, Blakeslee, Bridges, Morgan, Muller, Auerbach, to mention only a few.

2. Environmental Mutagenesis

The causative forces affecting variation and consequent possible disharmony or adaptive advantage to the organisms depending on selection pressures may be specific or diffuse and short term or long standing. The specific agencies and their cellular effects and consequences are well documented during the last 20 years. The same however cannot be said of the long-term, diffuse and complex mixtures affecting whole systems of organisms with probably more serious far-reaching consequences than apparent. We are now addressing ourselves to this question.

3. Pioneering Studies in the Field

While a global alarm to the danger of genetic risks from environmental sources was sounded by Bacq and Crow the initial results came from the background radiation in the monazite sands of Kerala coast of India where vegetation as well as marker species were screened for genetic impacts. An accidental mutation from gaseous emanation at Brookhaven were also found around that time. Klekowski subsequently found a native fern populations with interesting genetic changes as if under the influence of paper mill effluents in its riverine habitat. Fallow agriculture land with pesticide history also caused meiotic variations in some weeds. The mussel Mytilus was not only found to concentrate chemicals from marine surroundings but the tissue extracts have tested positive for mutagenicity. Thus the pioneering studies on the exogenous forces of variation related to the mutagenicity of background radiations, air pollution, water pollution and pesticides. And most of the work involved both plants and animals

4. In Situ Mutagenesis

Considerable literature has accumulated on the mutagenicity of various habitats, ambient environment and some sources of human subsistence. Mutagenic manifestations have also been monitored in plants and animals from mid 1970s as indicated above. As a result, a summary of onsite genotoxicity to plants was made by Grant and later by Ma who labeled the whole subject as in situ mutagenesis. Sharma however called the agents as Urban mutagens implying that these new mutagenic forces were created by the urban forces of consumerism-oriented development of essentially vibrant and resourceful ecosystems which characterize all rural situations. He further suggested several approaches of study in this issue and conceptualised the entire issue of monitoring approaches. Unfortunately the same has not been done for animals despite many reports on in situ effects, a few model assays and a review on fishes. Some inputs in this direction are naturally warranted.

The objectives of this review are therefore (1) to compile and summarize the entire in situ work on plants and animals and (2) conceptualize the monitoring approaches using both plants and animals depending on the ecology.

B. ENVIRONMENTAL FORCES

The causative forces of genetic change, unleashed by human activity pervade all the habitats of biota. These were discovered from different places in the world. All these are detailed briefly here - toxicants, habitats and regions

1. Toxicants

The polluting agents in various habitats are fairly large in variety. Some of them can be chemically classified while others can only be generalized based on their sources. A perusal of the listings suggests that they may be gases, solids or liquids, sometimes any two or all the three of these may occur together. Radiations are also prevalent in some habitats. The sources are also as wide as the chlorinated drinking water to radioactive wastes. A tentative list of about 20 classes of chemical groups has been culled from the literature. Similarly, as many were identified as sources of toxic emissions including 4 types of radiation sources. It should however be pointed out that only, some of these are known to cause genetic effects. Others can only be suspected until more evidences come forth than the current *prima facie* information suggests.

2. **Habitats**

The above pollutants are diffuse and widespread in the ambient environment in which organisms may dwell or pass through. Consequently they are subject to pollutant effects. A perusal of the monitored habitats indicates nearly 15 specific habitats receiving chemical and/or radioactive pollutants. These habitats vary widely ranging from predominant rice paddies and drinking water to rare sands of radioactivity. About 8 general habitats have been enumerated accounting for the above 15 specific regions. All these sometimes are over shadowed by a few rare occurrences of chemical and radiation accidents. Thus a great variety of pollutants, both chemical and radioactive, encompass a very wide range of habitats of plants and animals. However, genotoxicity was not reported from all these, but cannot be ruled out as per the current consciousness

3. Geography

The pollutants and habitats mentioned above exist in some definite places on the world map. A survey of these indicates the relative awarenesses and efforts by various countries. Thus the literature tells that the most of the work has been carried out in U.S.A. followed by Canada, U.K., Sweden, and Germany, with trickles from India and China, etc. A lot of places in the world are yet to be of investigated. However even these few places numbering about 40, cover coastlines, rivers, agricultural lands, estuaries, harbours, seawage and radioactive sources.

A perusal of the above information gives a false impression that a very great variety of pollutants, habitats and their locations across the globe have been monitored for genotoxicity. A close scrutiny reveals that all these are merely representative examples of pollutions and effects. Genotoxicity was *sensu stricto,* very little investigated in each category. Only a beginning is just made. This can be better understood by examining the actual work carried out.

C. MONITORING SYSTEMS - PLANTS

1. Systems Assayed

Habitats receiving varieties of pollutants have been subjects of assay by plants for a long time. These plants range from thallose algae to highly standardised higher plant systems. Thus several algae, a few mosses some pteridophytes and about 30 flowering plants native to the affected habitats and a few standardised test systems were used as monitors of habitat quality. These plants were both terrestrial as well as aquatic including marine waters. They have been receiving pollutants, through air, water and soil for different durations of time and responding in their characteristic way.

2. Endpoints Monitored

The endpoints scored in the higher plants were micronuclei, clastogenic and aneugenic manifestations in mitosis and meiosis, pollen fertility, pollen morphology and staminal hair mutations. Gametophytic abnormalities were scored in pteridophytes. However many studies have limited themselves to physiological endpoints suggesting accumulation of xenobiotics or their lasting impacts

3. Initial Studies

The pioneering work on in situ genotoxicity came from Gopal-Iyengar's group from BARC, Bombay who investigated the radioactive monazite sands of the Kerala coast. They found that several local plants were showing meiotic variations and Tradescantia has responded with staminal hair mutations. Almost at the same time found mutations in Tradescantia clones accidentally exposed to gaseous emissions from the power plant at Brookaven National Laboratory.

In 1973 Klekowski found apparently spontaneous mutation loads in the populations of the fern Osmunda. The subsequent works of this group has revealed a wide range of chromosomal and genetic changes in Osmunda and Mattuccea in response to pollutants from paper mill and power plant. Subsequently a fallow agricultural land in Quebec with pesticide history was found to harbour weeds with meotic variations.

The initial studies, thus, are quite varied. Two of these dealt with native plants on terrestrial sites while the third one also involved a native plant but under the stress of aquatic pollution. The possibility that monitoring systems could be scouted from native vegetation has thus been indicated. The other two pertained to a standard system - one discovering accidental air pollution and the other - an unsuspected habitat radioactivity. Thus the pioneering works focused the range of problems which have to be encountered in the real world situations.

4. Terrestrial Habitats

The terrestrial habitats screened for genotoxicity using plants are quite varied. On one hand there are geological varieties such as radioactive monazite sands of Karalla in India where both Trades anta monitor as well as native plants were employed. On the other extreme are the wastelands receiving effluents occasionally affecting the local plants by their solid deposits. Bikini islands received radioactive fall out due to atomic tests. These sands tested positive for mutagenicity in Tradescantia staminal hair system. Native plants on soils under the stress of pesticide, sewage and polymer effluents were screened for meiotic abnormalities and the consequent pollen sterility. While some plants have responded to these stresses, others did not. Sewage sludge, wastes from industry and water treatment plants and effluents from Smelters and Petrochemical factories have amended the soils in some places. All these were screened successfully using Zea, Tradescantia, Allium and Glycine. It is interesting to note that all habitats were found to be relatively genotoxic.

5. Aquatic Habitats

The pioneering and extensive work of Klekowski on Osmunda and Mattuccea, naturally growing on edges of rivers polluted by effluents from paper mill and power plant has already been mentioned. Laboratory effluents caused abnormal meiosis in Commelina. The well waters supposed to be under pesticide stress were screened in Tradescantia-MCN system while sewage in India and river waters in Sweden polluted by effluents from Herbicide factory and petrochemical complexes were assayed in Allium test. Waters in Italy and China were tested in Vicia-MCN assay and Trad-MCN assay respectively. All investigators found genotoxicity.

6. Aerial Situations

One of the early papers reported an accidental discovery of mutations in Tradescantia staminal hairs at Brookhaven National Laboratory premises. Interestingly, all tests on polluted air used Tradescantia Staminal hair monitoring system only. The surroundings of a Colorado dyke area and many regions in Alabama were screened as back as mid 1960s constituting perhaps the first report of in situ monitoring. But they were the heydays of concern about atomic radiations. Surroundings of nuclear power plants, Lead Smelter plants and other industrial complexes were also checked for mutagenic imports. By far, the most exhaustive and interesting work is of Schairer and Underbrink using a mobile monitoring vehicle with a lab at the ground level and a craning system to lift Trad-clones into air for exposures. Thus they screened a large number of sites in U.S.A. Unlike the terrestrial and aquatic habitats, all the air-sites were not found to be mutagenic.

C. MONITORING SYSTEMS - ANIMALS

1. Systems Assayed

The literature has shown that animals living in various habitats receiving pollutant stresses became objects of study comparatively more recently. Yet the works went accelerated, once began. The generic and common names suggest that Polychaetes, Arthropods, Molluscans, Echinoderms, Fishes, Birds, Amphibians and Mammals were used for screening genotoxic pollution in terrestrial, aquatic and aerial environments. Many of these were native to the polluted habitats while a few are standardized systems used as sentinels - or testers of transferred to the laboratory. The aquatic systems included both habitats marine and fresh water. Both in vivo and in vitro systems are used and in some cases methods were specially developed to meet the special exigencies posed by in situ genotoxins in the environment.

2. Endpoints Monitored

The endpoints scored were micronuclei, sister chromatid exchanges and other aberrations during mitosis and meiosis. Others are embryo lethality, larval deformities and neoplasms. Neoplasms assume great importance owing to the genetic basis of their origins - A conclusive evidence for the occurrence of genotoxins in the environment. Many studies have confined to isozymes and polymorphisms and both could have been used as genetic indicators.

3. Initial Studies

It is interesting to note that animal in situ genotoxicity was also found with radiations and in a native organism as in the case of plants. Polymorphism and cytogenetic exchanges were discovered in the larval salivary gland chromosomes of Chironomus in the white oak creek and lake receiving radioactive wastes from Oak Ridge National Laboratory, USA. But the lead was not taken up until late, when a mud minnow was found to have very suitable cytological features for genotoxicity testing. By this time the mussels were found to be accumulators of chemicals from environment. Taking advantage of this knowledge the concentrates from Mytilus from British waters were tested and found to be mutagenic. Taking the cue from the above, genotoxicity of Rhine water was tested. Since then the monitoring efforts increased the message having been picked up that natural environments harbour genotoxins. The studies were also highly varied - monitoring more habitats for pollution, scouting for suitable and sensitive organisms, standardising technical needs, and assaying for newer endpoints. As a result considerable lilterature has accumulated dealing with both aquatic as well as terrestrial habitats.

4. Terrestrial Habitats

Unspecified urban pollutions, air pollution, sewaged habitats and waste landfill sites were screened for genotoxic symptoms in the locally living animals or introduced monitors. The available data suggest that most sites are genotoxic to insects and rodents. The endpoints monitored were Sister Chromatid Exchange, Micronuclei, Chromosome aberrations, Abnormal Meiosis and Abnormal Sperms. A few cases of neoplasms were also reported.

5. Aquatic Habitats

The data on aquatic systems is nearly four fold than on terrestrial animals. The reasons are obvious - more pollution and more spread, and the animals would be more or less captive in the aquatic medium. Thus, harbours, estuaries, regional seas such as Baltic and Adriatic, rivers, lakes and ponds were screened by the locally available, or introduced monitors - or in the laboratory assays. Water surface layers and sediments were assayed by insects, polycheates, molluscans, frogs and fishes for genotoxicity manifestations such as chromosomal aberrations, sister chromatid exchanges, Micronuclei, Aneuploidy, Chromosomal exchanges, polymorphism and embryo deformities. It is significant that all habitats tested positive for genotoxicity. Many cases of neoplasms also have been reported.

6. **Aerial Habitats**

   Though air pollution is a known fact no attempts were made to screen birds. There is however some information on the cytogenetical variations in the birds of Japan and India. But it is also possible that water pollution may have indirect role to play here through food chain.

7. **Mussel Watch and Genotoxicity**

   Mussels are of world wide distribution - in the marine and fresh water systems of tropics, subtropics and temperate zones. They were found to accumulate chemicals from vicinity. This faculty was used to develop them as a monitoring system. By analysing mussels- shell as well as soft tissues one gets a window to the status of pollution in a particular habitat. This mussel watch program has been unveiled nearly a decade ago and is going from strength to strength If the mussels are also screened simultaneously for genotoxic symptoms one gets a picture of bioconcentration as well as its effect on the tolerating organism. The body extracts of these concentrators can also be tested in other organisms preferably a microbe. These assays combining concentration and testing can be called coupling assays – auto-coupling if genotoxicity is also screened in the same concentrator and allo- coupling if extracts are tested in another organisms. Further, bigger animals using molluscans as food can be tested both for biomagnification as well as genotoxicity. Therefore the mussel watch approach has a great potential not only to monitor aquatic pollution but also the genotoxicity and their passage upwards through food chain.

D. CONCEPTUALISATION

1. The Components

The components involved in the genotoxicity of the habitats or ecosystems are plants, animals, soil, water, air and human activities such as sewage, industrial emissions and human habitation. The climate - rainfall, temperature, sunlight and winds may also influence the interactions between the natural components of the ecosystems and the elements introduced into them by anthropogenic activities. The genotoxicity to plants and animals and the likelihood of similar dangers to humans ultimately depend upon the interactions between the constituent (biota) and additive (Pollutants) additive components of the ecosystem - a kind of residual impact. Inherent in this reasoning is the possibility that apparently innocuous stresses may compound to become obnoxious and vice versa from a genotoxicological point of view.

2. Monitoring Approaches

in situ

The ultimate objective is to monitor the final impact - a final diagnosis. The conventional approach would be to analyse the habitat for chemical and physical composition and assay their genotoxicity individually. This is not possible at the level of natural systems of association and interactions, which rule out simplistic approaches. Alternatively several approaches can be addressed to this question.

A. Denizen Approach

Firstly the naturally living biota can be screened for genetic responses - a kind of cumulative final impact after debiting or accounting all interactions. This may be called denizen approach. Only if an effect is seen, analysis of the habitat for toxic sources/chemicals can be taken up. But it is possible that the organisms may not possess reliable genetic markers or may have mechanisms to modify them or selectively concentrate and pass off into a distant organism through food chain, thus misleading the investigator. Therefore physiological, logistic, territorial and methodological problems may throw up false negatives, clouding judgments.

B. Sentinel Approach

To overcome the prospect of false negative data, organisms with delineated markers and sensitive to the environmental agents can be maintained in the putative genotoxic environment. These biological radars, through their well tailored responses, indicate the presence or absence of genotoxins. But there is a catch. May be it is too sensitive? May be it tells us only a passing phase in the long life of a habitat? May be it has enzymatic systems activating or inactivating certain chemicals. All these pose the problem of relevance of the data from sentinels to the real world situation in a habitat. To circumvent this snag a more direct method can be followed - the third option. *ex situ*

C. Habitat Sampling Approach

Here, the environmental fractions are brought to laboratory, maintained and screened in well established - test systems in vivo and in vitro. In other words, water, soil, sediment, surface microlayers and extracts of any of these can be tested in standardised plant and animal test systems. The problem here is one of ignoring the dynamics of these agents in the physical, chemical and biological milieu of the ecosystem.

Thus each approach has advantages and disadvantages. If all are put together in a single regime all questions are answered. In other words, all monitoring approaches had to be followed so as to account for all aspects of pollutant dynamics in nature. Thus the denizen approach covers cause and monitor together in the field, the sentinel approach takes the experimental monitor into the field while the habitat sampling approach brings the field into the experimentation. Both in vivo and in vitro conditions are taken care. Metabolic activation can be covered by incorporating the necessary step into the assaying protocols.

D. Bioconcentrator - Assay Coupling Approach

Despite all these, yet a 'danger' may lurk in the form of minute concentrations of some chemicals in the medium entering the biological materials with long latencies to accumulate enough to cause an effect. This effect genotoxic, among others, may be in the same organism or in its consumer. Thus an "element" of the habitat is concentrated in an "element" of the habitat biota to monitor whose toxicity a third "element" of the laboratory has to be used. One has to identify the bioconcentrating organisms, extract the chemical groups and then test them in any sensitive system in vivo or in vitro with or without metabolising enzymes. This is a "kind" of *in vivo* host mediated assay and Ames assay combined to monitor a difficult ecological situation.

The details of approaches to monitoring genotoxins in the habitats are known. The Denizen and Sentinel approaches are essentially in situ assays, while habitat sampling and concentration Assay coupling are ex situ assays. These monitoring approaches suggested here for animals are the same suggested for plants In the present study however we obtained evidences from the literature on the correctness of these approaches, both from plant as well as animal studies. All plant studies considered here easily fitted into the formats based on this concept Animals also fitted similarly. Additionally the data on neoplasms and concentrator-Assays also fitted well into the concept of monitoring approaches.

Thus we hereby claim to have paved the way for an orderly approach to genotoxicity problems in the habitats in preference to erratic ways and freelancing.

3. **Relevance Of Bioconcentration**

Animals and plants concentrate chemicals from the environment Animals can also concentrate through food chain. But all plants and animals cannot do these, only some can. Even these, can concentrate them to a point beyond which they tend to be toxic. It is possible that tissue levels of some substances are far greater than the ambient levels. Concentrates of such chemicals are important in genotoxicology. Ambient levels is not amenable to genotoxicity testing due to a minuteness of only concentrates can be tested. If the bioconcentration by organisms or tissues is selective, such a natural fractionation of pollutants render toxicity assay simpler and definitive.

4. **Relevance of Neoplasms**

There is a large body of literature on neoplasms a part of which is currently compiled. All these tend to suggest the occurrence of genotoxins in the habitat without identifying them. This implies, further, that other genetic consequences such as teratagenicity and heritable changes may also be taking place in these habitats, may be in some other organisms. Further more, it is possible that these genotoxins enter human systems through food chain, for, most of these results came from the edible species.

The significance of neoplasms in natural populations of animals is that they prove the existence of genotoxins or their precursors in the medium. Whether the agents are concentrated by direct intake or through biomagnification will be of significance to the predating species including man. These data are very relevant, for, neoplasms can be cited as one of the endpoints suggesting genotoxicity.

E. PERPSPECTIVES

1. Monitoring Needs

The first inference that can be drawn from the foregoing account is the necessity of monitoring in situ hazards following the conceptual framework to avoid erratic accumulation of scientific data. Then it would be seen that a paucity of test systems and methodology exists. A close scrutiny however assures that there are enough plant and animal species with suitable features for employment in in situ monitoring. By drawing from the available literature a few candidate plants and animals are suggested

There is however a greater need to identify the regions to be monitored. Taking India as an example, estuaries, river mouths, coastlines and wetlands are the first category to be studied . The second category relates to the urban regions where considerable industry is concentrated already. Both these together cover terrestrial, aquatic and aerial habitats of India.

In this connection, mention must be made of the relevance of mussel watch approach - if mussel populations are present in adequate levels. The approach consists of assaying the mussels for chemical contaminants as they are confirmed concentrators and testing these (i.e., auto -coupling) or their extracts (i.e., allo -coupling) for genotoxicity, will caution us if similar problems await human populations as well.

All river systems in India need similar assaying as there is no river in India that is not receiving effluents of some industry. In fact rivers, estuaries, coastlines, wetlands, urban systems of the developing countries need to be screened.

2. **Ecological Implications**

The bioconcentrations, genotoxicity and other deformities in the animals of various habitats, suggest that all these ultimately result in variations and polymorphism. But these variations may put the population to advantage or disadvantage *vis-a-vis* the environment. Coupled with the occurrence of neoplasms, what may be the ecological impact of these happenings - changing population densities, migrations, interactions, food chains? Will these ultimately result in genetic diversity and speciation or depletion and extinction of some species? It is possible that both may be happening in different species.

3. **Human Health Considerations**

Strictly speaking, concern for human health from polluted habitats constitutes an ecological implication of in situ genotoxicity. Human health may be influenced genetically by the same environmental forces that act on plants and animals. Thus the occurrence of neoplasms in animals may be similarly suggestive of similar end result in humans. If these animals are consumed by man, it is possible that the agents causing cancers persisting there in will enter human system. Thus the human society is stressed by the genotoxic agencies both ecologically in small measures and through consumerism in lesser measures.

F. Conclusion

Through the long lists of references, agents, habitats and species one might construe that remarkable work has been done on in situ genotoxicity. A close look reveals the contrary. The information is just enough to conceptualise the monitoring approaches and develop a model of genotoxin tracing and passage in the ecosystems. While there are enough examples to fit into the various monitoring approaches, and hazard hypotheses, considerable blanks await to be filled. Most references pertain to bioconcentrations without monitoring for genetic consequences. Many studies reported neoplasms without going to genetic roots. Obviously, a great deal of researches is awaited. Yet the very occurrence of neoplasms, in native animals, the phenomenon of bioconcentrations and reports of genotoxicity together constitute a multiple caution to the safety of human genome (and genotoxicity of plant and animal genomes) - all over the globe where economic development is taking place. A neat relationship with politics, economics, natural resources, ecology, pollution, toxicology and human ill health of irreversible kind devoid of warnings is established.

12

ECOSYSTEM MUTAGENS

REFERENCE: *Sharma CBSR (1992] Ecosystems harbour dynamic genotoxins. Concepts of monitoring and implications to human health and ecology of plants and animals, Lecture at the MSS Research Foundation* **Chennai** *to a National Work Shop -*

A. INTRODUCTION

Most ecosystems are under some stress of pollution either directly or indirectly -air, water, soil and food chain etc. These ecosystems support plants and animals and connected to man either directly or indirectly. The stresses, on these biota among others could be genetic. The chemicals may be in small quantities, in mixtures and therefore, dynamic, complicating an understanding of casual relationships. 1st ascertaining whether or not ecosystems harbour genotoxic agents is essential for assuring a reasonable health status to humans.

B. GENETIC EFFECTS

Mutagenicity Clastogenicity and aneugenicity of chemicals and radiations in plants, animals and humans were established between 1927 and 1967. In humans. birth defects, cancers and heritable diseases are caused as a result The exposures may be through accidents, ecology, occupation, consumption and usage Other organisms are also exposed through accidents, ecology and food chain.

C. CONCEPTUALISATION OF MONITORING

The information available so far is both from animals and plants from aquatic and terrestrial habitats. Most of these are however, circumstantial. Based on the inferences from these highly scattered literature the monitoring approaches can be conceptualised as follows- which alone matter when the agents are diffuse and dynamic:

1. ***in situ* MONITORING**
   (a) Denizen approach *(in vivo)*
   (b) Sentinel approach *(in vivo)*
2. ***ex situ* MONITORING**
   (c) Habitat Sampling (*in vivo & in vitro*)
   (d) Bioconcentrator - Assay
      (i) Allo coupling (*in vivo*)
      (ii) Auto coupling (*in vivo*)*

D. MONITORS AND RESPONSES

Summaries of genotoxic responses in plants and animals in various ecosystems are provided annexures as per the above scheme of conceptualisation. The endpoints scored were: Sister Chromatid Exchanges, Micronuclei. Meiotic and Mitotic aberrations, Aneuploidy Embryo Deformities and Neoplasms in animals and Meiotic and Mitotic Aberrations, Micronuclei, Specific mutations at *waxy* and *pink* loci in plants. The plants assayed were native plants,

Alluim, *Vicia*. *Tradescantia*, *Zea* and a few ferns. Animals were mostly molluscans and fishes. Others are Insects, Polychaetes, Rodents and Birds. The literature has about 26 references on plants and 31 on animals suggesting the comparatively recent origins of this discipline connecting pollution toxicology, genetics taxonomy and ecology. A high light of these approaches is that some organisms concentrate chemicals and exhibit genotoxicity as a response !

E. PERSPECTIVES

1. Human Health Implications

(a) Gene based diseases may have ecosystem origins.

(b) Vectors and pathogens may undergo mutations resulting in new varieties of both.

(c) The above can aid in developing new perspectives on epidemiology and preventive medicine.

2. Ecological Implications

(a) Spontaneous mutations reported in literature may be a myth

(b) Genetic death and polymorphism may occur in a large scale with increasing pollution.

(c) Genetic changes in humans, vectors, pathogens pests, weeds, parasites may assume an epidemic form endangering human welfare both directly and indirectly.

3. Imperatives

(a) A reconnaisance of all possible genetic Homoclimes and a concurrent search for new monitors and better techniques are warranted.

(b) Linking the sources - agents - habitats - Organisms - Effects - Humans may be desirable to develop management strategies.

(c) The international mussel watch programme may be more, genotoxicity oriented and more widely implemented in India.

13

ENVIRONMENTAL MUTAGENESIS IN INDIA

REFERENCE: *C.B.S.R. Sharma, N. Pannerselvam & Arumikkili, A Survey of Environmental Genetic Toxicology in India Environmental Mutagen Society of India, 1985/69p, EMSI did no update this so far.*

The importance of genetic insults from environmental agents to human was recognized during the 1960s and paved the way 'for the, organization of, the Environmental Mutagen Society in U.S.A. Concerted efforts following this have resulted in the screening of environmental agen1s by using appropriate assay systems drawn from the preexisting literature in genetics. This message reached India when the environmental Mutagen Society of India (EMSI) was given birth in 1975, chiefly due to the efforts of scientists at the Bhabha Atomic Research (BARC) led by Dr. K. Sundaram. Like the fraternal bodies in a few countries, EMSI has been striving to organize efforts to ultimately provide the public with information on the genetic toxicity of their environment, from where there is no recourse to normalcy; once set in.

HISTORICAL ASPECTS

While the investigations on cellular responses to chemicals were first carried out during 1890s, well documented results appeared only around 1940. These included such chemical groups as pesticides, alkaloids, drugs and even routine laboratory reagents involving celebrities such as Kostoff, Eigsti. Auerbach, Robson, Oehlkers and Rappaport. lt is interesting to note that one of the early studies on pesticides by Kostoff was published in an Indian journal, Current Science, A survey of the Indian scene in this discipline of prime human concern may be worth while.

EARLY INVESTIGATIONS

The earliest papers in India were by the great botanist, Prof. Mehra, In 1946-9 he studied the effects of colchicine and sulfanilamide in Ephedra pollen. He also pioneered similar investigations on pteridophyta which today command great appeal for *in situ* employment. Spermatozoan abnormalities in fern prothalli in response to chemical stresses were also assayed by him for the first time Incidentally, the first review in this field was also by him. During this period foundations were also laid in some allied areas. Responses to Gammexane, Ethylene glycol, Carbon tetrachloride, Vegetable oils insecticides Urethane were documented in systems such as *Corchorus, Vicia, Arabidopsis, Allium; Drosophila, Poecilocera* by Rao (NS, SRV) Seshachar, Bhattacharya, (s) Swaminathan, Natarajan and Kundu. Prof. Seshachar was perhaps the first to recognize chromosome segregation as a source of genetic errors.

Works of Prof. A K. Sharma from mid fifties are a class by themselves dealing with a wide variety of chemicals, Which were major trend setters in the cytogenetical research in India, ever since. During the close of this period came the most significant contribution i.e. the genetic toxicology of the monozite belt in Kerala using on site vegetation and *Tradescantia* staminal hair system. This perhaps is the first *in situ* investigation in the world, for the kind of which much stress is laid presently. Publications from Delhi and Hyderabad have just begun to appear from Natarajan, Kaul, Kesavan. Rao, Reddy, Subramanyam etc. Also appeared at this time was a review by Sharma (CBSR) on plant systems. Investigations being fragmentary, It is apparent that the progress has been devoid of direction.

LATER STUDIES

When the EMSI was organised as an affiliate of the International Association of the Environmental Mutagen Societies, efforts were made to provide a direction to such investigations. Annual Conferences, Seminars and Workshops were organised and a Newsletter distributed periodically. As a result scores of workers all over India are today engaged in assying genetic stresses in model systems. Three major centres can be identified currently which however had independent origins.

MAJOR GROUPS

The foremost among such centres is the one led by Dr. Sundaram at BARC whose work was recently reviewed in the Newsletter of the EMSI. This group has not only EMSI but also nurtured it to be one of the best of India. A large number of agents-drugs, food colours, dyes, industrial chemicals, radiations etc were studied by them in microbial and mammalian systems. No wonder that the Department of Atomic Energy was the first agency to recognise the genetic toxicology as an area for funding.

Osmania University is system intensive possessing expertise in plant, microbial in *vivo* and *in virto* systems. Most publications of professors Nizam, Reddy(OS and GM), Subramanyam, Rao, Ahuja, Polasa and their colleagues have so far dealt with drugs, natural products, pesticides and others. Though their interests are wide ranging involving an equally wide variety of endpoints, more emphatic studies were made on drugs.

The most remarkable is the current pace of agent-intensive activities at the department of Botany of Calcutta University led by Prof. AK. Sharma, Prof. A. Sharma and Dr. G. Talukder. Having initiated chemical screening for earlier to the pre - EMSI period, the centre is on full steam with scores of workers screening a great variety of agents such as drugs, naturals, industrials, pesticides in plant and mammalian *in vivo* system

OTHER GROUPS

Of lesser magnitude and pervasion, but often greater in depth are a few smaller groups spread across universities and research establishments. Significant contributions were made from these locales mostly specialising in specific systems and specific chemical classes. Among Such are the groups led by Kaul, Sharma, Dikshit, Grover, Sobti. Singh, Nigam,

Kesavan. Marimuthu. Nagarajan, Gopinath, Bhide, Bhattacharjee, Vaidya and Manna from Jammu, Madurai, Lucknow, Amritsar, Chandigarh, Ahmedabad, Varanasi, Madras, Delhi, Bombay, Calcutta, Poona and Kalyani.

Only some centres of activity are presently described though there are many individuals engaged in genetic toxicology labours. Therefore a list of Indian work is included. It may be noticed that this bibliography is pregnant with indications that more active groups are likely to emerge in future.

TRENDS

The literature search is very suggestive. A perusal of the publications following this write-up and the appendices on journals, systems and agents reveal past achievements- some of which are wasted and current trends - both disquieting and healthy.

Some of the early achievements in India pertain to the (a) use of fern systems and spermatozoan abnormalities, (b) early recognition and use of systems such as *Drosophila*, *Poecilocera*, *Arabidopsis*, *Lycopersicon* (c) usage of on site vegetation and *Tradescantia* to monitor in situ radioactivity for the first time in the world and (d) establishing micro-nuclei and segregational errors in plants as endpoints incidently by some zoologists! (Mehra, Bhattachrya, Jain, Rao, Gopal - Ayengar, Seshachar, Shah and their associates.)

More recently, pesticides received preferential attention from groups led by Kaul, Singh, Sharma and Dikshit employing plant and mammalian systems. Drugs were mostly screened by the groups from Osmania, Calcutta and BARC, the latter infact covering a whole array of agents that human populations are likely to be under exposure. Naturals were stressed by Dr. Subramanyam and the Calcutta team.

Much however cannot be said concerning our contributions to the development of test systems inspite of historically significant publications. The great tradition in plant cytogenetics has contributed for the wider expertise in plant systems copared to that in others. Unfortunately this was not exploited appropriately and mammalian expertise therefore appear to be deeper, stronger and more reputed. Yet it may be pertinent to remark that some of the foremost genetic toxicologists in the West who are Indians happen to be from botanical background.

DISTURBING ASPECTS

The exemplary work carried out by the pioneers seem to have been ignored for long. Otherwise it is difficult to explain the disrepute some of the later works have acquired. Even today, two disturbing trends are discernible from the literature which may upset the efforts to enhance the credibility and utility of research labours in genetic toxicology.

(a) Even after the concepts of environmental genetic toxicology were developed where criteria for the usage of test systems, protocols, dosage, patterns are well established, sampling. scoring and inferences continue to be erroneous in many papers, And this unhealthy trend seem to attract many a new crop of workers! Consequently, over the years there appears to be no significant increase in the reproducibility of the work from certain quarters. This hopefully may be a passing phase. Nevertheless should be noted that many papers on irradiated foods, pesticides and drugs, especially dealing with mammalian, and plant test systems are increasingly cited in the western works, considering this as a measure of our credibility.

(b) Same work often appeared in different journals with different titles. Some chemicals are tested many times ignoring precedents even from India. Essentially single piece of work is often split into several publications. Data on a single chemical from different systems and from a single system on different chemicals were published in multiple forms. Some publications do not reveal the identity of either the test system or the test agents. Reasons for such a state of affairs are not far to seek. Some of the journals are of doubtful circulation, reputation and even integrity. Tendency of publishing unreviewed papers on a large scale may also be another reason. Excessive haste in putting into press as many titles as possible, dealing with the widest spectrum of agents to gain historical precedences, if not reckoning, is also a factor in the gradual erosion of standards in the mutagenicity research. Often this develops into a notoriety!

In order to gain in credibility something has to be done to stem these disquieting trends. However it should .be noted that the scenario is not that bleak, but is replete with major contributions. A few selections are detailed.

ACHIEVEMENTS

Based on the nature of effects, that some chemicals can be used in chromosome techniques was first recognised by

Prof. Sharma. Also that growth substances and natural compounds may have a role in the mutagenicity was also recognised by him. The efforts by him, his wife and their group in this and allied fields were unparalled in the history of Indian science in their sheer sweep. No wonder that the book by them on chromosome technique carrying information on chemical effects has become a major reference source all over the world. And scores of workers in India try to emulate them in their research efforts and scientific ethos.

Relationship between cell cycle and time schedule of harvesting cells for aberration scoring, the importance of micronuclei (Shah and Rao), interactions between radiations and chemicals (Kesavan) and between chemicals and DNA (Podder), utility of the Indian Muntjac as a test system (Sharma. T) and chiasmata as a responsive endpoint (Sharma) were among the major areas of contribution from Indian workers. That the associations of chromosomes during segregation could be influenced, which has relevance to Centromere synchrony and turbagenicity, was discovered by Dr. Natarajan long ago. Cellular pictures in his publications and in those of Prof. Mehra and Dr. Subramanyam are probably of the best of standards. It is only to be expected that some reviews should also have appeared.

The contributions from the workers at the Bhabha Atomic Research Centre are exemplary. Internationally reputed and nationally torch bearing work on aspects such as *in situ* radioactivity irradiated foods, drugs, dyes, risk estimates, radiation equivalence of chemicals and on systems dealing with Ame's assay, Barley, *Arabidopsis*, *Tradescantia*, Micronuclei, dominant lethal test, clastogeny assays *in vivo* and *in vitro* and on all aspects of yeast were done have been continuing here. [Sundaram, Murty, Chauhan, Bhat, Bhattacharya, Chaubey, Aiyar, Notani, Gopal-Ayengar, Rao, Bhatia, Pradhan, Bhattacharjee, Aravindakshan].

PRESPECTIVES

With the historical experience and the established expertise behind us attention could now be selective so as to accelerate the natural process of development. (1) Identification of hazard sources (2) building up of relevant data bases and (3) generating an ethos of credibility in our efforts and reproducibility in the investigations are thought to be among most primary. Among the issues, the following are suggested consideration.

(a) The *in situ* situations such as sewage, effluents, ambient air, waste-fills, occupational atmosphere, home-stead environment have to be investigated for their composition both in the urban as well as rural human settlements. It is indeed in the home front that most pollution seem to occur!

(b) A more imponderable situation is posed by the food chains. Genotoxins can pass through vegetables, mllk, meat and the processed foods. Many consumables are often adulterated adding insults to injuries. Further, the consumables are increasingly reared on effluents, sewage and other wastes, sometimes surreptiously, compounding the hazards.

(c) India is famous for the great variety of native medicines, culinary delicacies and gastronomic recipes. In view of their complexity and considering the large populations involved, investigations in this virgin field may be worth while.

(d) Various biota of the agro ecosystem may experience mutations in an unprecedented scale from pesticides and their modified versions, given the realities of modern agriculture, upsetting the existing balance of forces. Such a *mise en scene* has seeds for disastrous consequences due to mutation pollution. An ecosystem approach in genetic toxicology of pesticides is therefore warranted.

CONCLUSION

A co-ordinated programme appears to be necessary for each kind of hazard-source. Thus Air, Water, Consumables, Agrochemicals etc., can be investigated by teams representing the various test systems coordinated by one among them. Assistance of all agencies relevent to environment has to be obtained. This new aspect was found only as a passing reference in a recent hand out of the Department of Science and Technology! Workshops to train personnel in the methodology are essential and happily there were two so far iii 1974 and 1982 at BARC. Since credible schools are operating in this country with expertise in various systems one can take effective steps to minimize fragmentary reports based on expedient research. If there is one area where team efforts are necessary to avoid costs, duplication, and waste of man power, it is probably here!

An estimation of and investigation into, the genotoxic threats in India through consumption, usage, occupation and/or ecological routes may be best served by an objective assessment of our own strength to gauge the enormity of this problem. If this thought is shared by others, the pains of our bibliographical labours would be without a hangover.

14

GENEPIDEMICS & GENEPANDEMICS IN THE ECOSYSTEMS AND HUMAN SOCIETIES

A Trans-Disciplinary Approach - Environment and Health - Beyond. ROCHS-NIOH-ICMR-Complex, Bangalore, 16-19, September, 2009.

C.B.S.R. Sharma

An Ominous Possibility emerging from a Reconsideration of Darwinism in its 150th year

A. INTRODUCTION

1. Changes in the genetic systems of organisms including man are caused by chemicals and radiations. These mutations multiply with time and with each generation, horizontally due to the continuation of exposure to such agents and vertically due to the possibility of transmission. In human populations they cause cancers, birth defects, heritable diseases and minor mutations with varying detriments to human performances. In the ecosystems, they cause varieties of mutations in plants, animals and microbes. They include food animals, vectors, pathogens, pests, parasites, infectious as we as useful microbes.

Individually or collectively they impact human welfare adversely.

2. Such consequences may be termed Genepidemics and Genepandemics. The former are limited to populations sharing same environmental components in a limited space say around polluted habitations. The Later could occur any where in the world where the populations are exposed to the same kind of chemicals or radiations.
3. Both the above mostly occur in the developing countries due to relative illiteracy, awareness, knowledge and higher levels of needs and wants.

B. PREMISES

The evidences are considerable - direct, indirect, circumstantial from human societies and ecosystems- both managed and natural, terrestrial and aquatic. Though we tried to integrate all these into a new field of investigations 15 years ago, combining ecology and genetic toxicology, when I first suggested this possibility in a conference in Malaysia, it could not happen. This remained a tragedy of my scientific pursuits, all my life.

C. EVIDENCES FROM FACTS AND PORTENTS [CIRCA, 1995]

1. Human Ecosystems

(a) Diagnosis and drug Administration: Resistance caused among the causative pathogens/parasites to various drugs can similarly be suspected to be one of the side effects of the drug. And some drugs are known to be genotoxins.

(b) Resistance of vectors such as mosquitoes to pesticides can also arise from the sub lethal action of toxic chemicals either because it is adulterated or it is not uniformly applied or the instructions of the pesticide companies are not followed *in toto.*

(c) Scenarios: Birth defects varied from 0.5 to 2/5 % Mental Illnesses were around1% 4000 syndromes were already recorded. 25% of people were projected to be afflicted by Cancers ultimately and minor ailments may cause detriment to varying degrees. 35 species of Anopheles and 41 species of Culex became resistant to Pesticides. Parasites causing Malaria and Shistosomiasis also developed resistance to drugs

2. **Agroecosystems**

(a) Agro-ecosystems: Resistance caused among pathogens and pests such as fungi, insects, nematodes and others to fungicides, insecticides and other pesticides could be mostly due to genetic effects of sub-lethal doses.

(b) Scenario: Resistance was recorded in about 370 insect pests to 9 pesticides, including. Those of Cotton, Rice etc. 7 Rodent species became resistant to Rodenticides. Mutations were noticed in rhizobia blue green algae, weeds and Pollinators

3. **Polluted Ecosystems**

(a) Aquatic Ecosystems: Pollutions enter animals which could move along the food chain to human societies– whose consequences cannot be easily discerned nor monitored

(b) Trans-Boundary Ecosystems: The most ominous and consequential aspect pertains to Ornithology. Birds cover aquatic, terrestrial and aerial environments of local, regional and global pervasions and are recipients of an un-definable array of chemicals –gaseous, liquid and solid particles. The regions they visit temporarily as migrants can be at the receiving end of alarming developments, such as dropping of pathogens through Guano. [This is other than such acts of using them for aerial photography of sensitive installations before blowing, as in Bharatpur]

(c) Scenario: 90% of studies on animals - crustaceans, polychaetes, amphibians and rodents demonstrated birth defects in their natural habitats. Many animals exhibited cancers in response to the aquatic media. Several plants and animals turned out to be bioconcentrators. A rare case of speciation in Molluscs within a few generations in two very distinct geographical zones was also recorded in the British West Coast and Californian Coast in response to pollution.

All the above variations are due to Mutations in soma or germ-line. And all these are projected to increase in time. Further all those recorded in developed countries are expected to repeat in the developing countries which opted for the capitalistic mode of development in the name of Globalisation.

Genetic Toxicology involving agent-induced genetic changes and their consequences of muatagenicity, carcinogenicity and teratogenicity to human systems now becomes Environmental Genotoxicology - a hybrid of Ecology and Genetic Toxicology embracing other organisms, their ecosystems their ecology, food chains, movements, etc. This naturally takes us into the fascinating world of nature.

Incidentally this is the 200th anniversary year of Darwin's birth and 150th anniversary of the birth of Darwinism and is termed Darwin year in the west. Wide ranging celebrations through Seminars on the futuristic implications of Darwinism from diversity and extinctions to Social Darwinism and exclusions of people.

D. DARWINIAN IMPLICATIONS

1. It is well known that **biodiversity** has been dwindling due to a variety of reasons. One of them is the pollution of air, water and soil by chemicals. Pollutants not only cause variations or mutations in the extant biota but also select the suitable ones. The ultimate impact of these forces is the acceleration of genetic and ecological diversity in the gradually decreasing taxonomic diversity - an aspect that is not taken seriously yet, rather strange! This situation is diametrically opposed to the scenario existied upto post-colonial times. What are the implications of such a changes which the Developed countries only are conscious?
2. It is not widely appreciated that the **stability of GM Crops** will depend on the basic genetic system of the crop and methodology employed for its modification. In the former allelic natures-multiples and pleiotropy,

polygeny, position effects, trasposons, sex determining mechanisms and such special mechanisms are to be considered. The method of evolving any GM crop will be a trade secret. Agroclimatic conditions are another important matter. The concern here is that pollutions causing mutations in other biota could cause the same in GM crops, affecting their stability and therefore, yields. Food security is directly under threat.

Further all these may happen in a defined region, as an epidemic or in several smaller enclaves sporadically, as a pandemic, both confusing and confounding every one!

E. DETRIMENT INDEX

The concept of detriment index was developed by an Indian born Dutch scientist Sankaranarayanan and a Hungarian Scientist Czizel based on the Hungarian Health Statistics, probably the best maintained in the world. According to this concept every genotoxic effect restricts the normal life of the victim by bestowing its characteristic effects. To this extent the individual suffers a detriment, i.e.loss of work, productivity, self improvement and ability to contribute to common development.

The detriment is expressed in three ways. 1. Years of lost life [LL]. 2. Potentially impaired life [PIL] and 3. Actually impaired life [AIL]. These indicators are derived from the frequencies of 26 common multifactorial diseases [6500 / 100,000]. These diseases are classified as very severe, moderately severe and less severe. With the aid of some omplicated statistical and clinical analysis it was estimated that these diseases caused about 2700 years of LL, 96,000 years of PIL and 5800 years of AIL per 100,000 individuals of Hungarians. The multifactorial diseases seem to cause greater detriment than the mendelian ones.

F. FINAL DIAGNOSIS

One must remember that Gene Epidemic portents would mean

(a) at least 10% of gene abnormalities in the human population which will increase with each generation and

(b) 10-25% of cancer subjugation and ultimate mortality through unprecendented suffering.

Some estimates reach up to 60%, both put together! All these constitute an enormous social burden on the already fragile economic fabric of the third world. A revolutionary change for the worse in the health condition of mankind in the third world is clearly predicted. In the least, the issue is highly emotive. With populations on the rise, food supplies faltering, the spectre is indeed frightening and a shock or a disaster in this direction is on the cards. Pestilence stealth in darkness!

If Malthus and Darwin's postulates on population were averted by modern science, the products of the same science has portents to unleash epidemics of genetic origin and prove these prophets tellingly true unless intervened by a nuclear holocaust in between.

15

FINAL RECKONING

RADIOMIMETY TO LIFE-STYLE MUTAGENS TO ENVIRON - MUTAGENS TO ECOSYSTEM MUTAGENS TO GENE - EPIDEMICS

I have been trying to understand what I was doing soon after I started working for my Doctorate; I did not succeed; but efforts continued, yet no success. But this pattern continued, probably even now. During 1966 I realized that I was not correct, after spending some time at IARI, New Delhi and listening some scientists during Mendel Centenary Symposium – which was filled with MMS and EMS as the latest buzz words. I came back and jumped on to plant products as they are more prominent and abundant than any, Prof. Santra brought me Vinca Alkaloids from Eli Lily's when he visited MIT. So if they have Sulfonates, I got Vinca alkaloids then most potent anti-neoplastic drugs. If sulphonates produced chromosome breaks and mutations, alkaloids caused the former and the latter, unknown! So near and yet so far. These Implications in mimicking radiations were understood in 1971 only, following the publication of my review on Chemicals and Cell division in JSIR. It took 8 long years to understand my own work, Lucky students get guided by their mentors. Some need to struggle. This also is a pattern that runs in some lives.

Thus a quest for chemicals that function like radiations in genetic effects has been going on where Kihlman, Biesele were in the forefront. I too added my foot which enabled me to discover that mutagens were suspected to be in our lifestyles – the field of Genetic Toxicology was born Some of these mutagens were pesticides already made famous by Rachel Carson. Due to their genetic toxicity, insects, fungi, weeds, etc developed of resistance to pesticides. Since agriculture is a balanced ecosystem, its instability was feared. Back to life styles, more hazardous chemicals were discovered to be mutagens – necessitating a modification of the central dogma. We realized soon that pollutions are increasing endangering the poorer people and the pollutants are not the same as the life-style examples. Also it is found the entry routes and distances too vary. Therefore an ecosystem approach needed to be adopted – labs needed to companion with fields. Candidates for assaying emerged from native species; the entire understanding begins to undergo a sea-change. Since most genetic changes are deleterious, and likely to be in large proportions and spread freely, the term GENE-EPIDEMICS could be appropriate. Each of the above stages took 5-10 years each. So the time passed away in struggles, in discovering the real epidemics to ourselves.

In every summation and assessment history is inevitable. Thus "Mutation", a big variation, was first heard 120 years ago from De Vries. Muller tested Roentgen's Rays and found them mutagenic 20 years later. During 2nd World War, a constituent of a Chemical weapon, Mustard Gas, Methyl isocyanate, was found to be mutagenic by Charlotte Auerbach. This classified information saw the light in early '50s. when the biological impacts of atomic radiations were assessed by the Crow's Committee.

Thus both radiations and chemicals were confirmed to bring about heritable changes in biota, Mutations. Researches increased on the nature of mutations in different organisms and to search for chemicals which can cause mutations just as the radiations, the radio-mimetics. This is around the mid '60s. While the former was mostly concentrated on lower organisms such as bacteria and fungi, the latter were exclusively on higher plants and animals due to their relevance to carcinogenesis in humans and yields' improvement in crop plants.

I heard of the developments and about EMS, Ethyl Methane Sulphonate at Indian Agricultural Research Institute's Genetics Department, where I used to spend all vacations in Dr. M S Swaminathan's Lab to work on my Doctoral thesis. Consequently my work on the effects of Vinca alkaloids became 50% of my thesis ! By 1969 a Society on Genetic Toxins was formed in USA, called EMS, because Ethyl Methane Sulphonate [EMS] was the most popular and potent mutagen in research and applications at that time. Otherwise it was more a Genetic Toxicology than Environmental Mutagenesis. This was the time when the world was witnessing disastrous events in its environment. By the time EMSI was initiated in 1975, basic concepts were being agreed to yield a conceptual framework. Chemicals in direct use by humans were the focus – the life styles. This became the central dogma.

Almost simultaneously and there after, focuses were centering on Pesticides, which are designed to be toxic. The primacy was shifted from life styles to Agro-ecosystems. This is for 2 reasons. Crop yields on which human lives depend may be reduced due to undesirable resistances to pests and sensitivities of crops, genetically. Further such pesticides could pass to human systems through food chain becoming a life-style genotoxin - as residues or metabolites.

Pesticides would mean not merely residues but also their degraded or modified substances. Food chains need not be short but substantially long, thus paving the way to any other substances in the related systems. Thus the agents, routes and impacts become complex. All these force revising the earlier premises and other related aspects consequently. And more systems and pollutants became relevant to genetic toxicology. The subject now becomes truly Environmental Mutagenesis! Thus the theory of Gene-Epidemics had to make an entry!

In situ studies further changed the scenario to greater complexity but nearer to the total truth. Economies becoming capitalistic, developing countries turning globalized, production, consumption, buy and sell syndrome becoming our new Gods during the last 3 decades. Consequently the number of chemicals released into air, water and food stuffs became multifold. They could genetically affect not only humans, but any organism– small or big – resulting in a swarm of mutants, These could impact any other organism – importantly, the humans. This is unforeseen but unpreventable – bolts from the above. The present Covid is one like this - a Mutant Virus about which most did not know anything until human health is impaired. Similar ones may be in the waiting to appear in due course. – the Coming Plagues, Laurie Garret termed!

Thus it would appear that Environmental Mutagenesis has greatly expanded in 50 years – first limiting to Human Societies, then Agrosystems, and appeared to end in Polluted sites. But now it appears that the spread is all-encompassing all environments, all food chains, and all organisms; further some of these organisms damaging the humans- truly epidemic and pandemic ultimately. A whole ominous subject is awaiting for detailed studies with an Ecosystem Approach, for, the Gene- Epidemics have that much expanse and depth to probe.

A. KNOWLEDGE IN NUT SHELL

DOCUMENT - 1
WHAT IS TOXICOLOGY

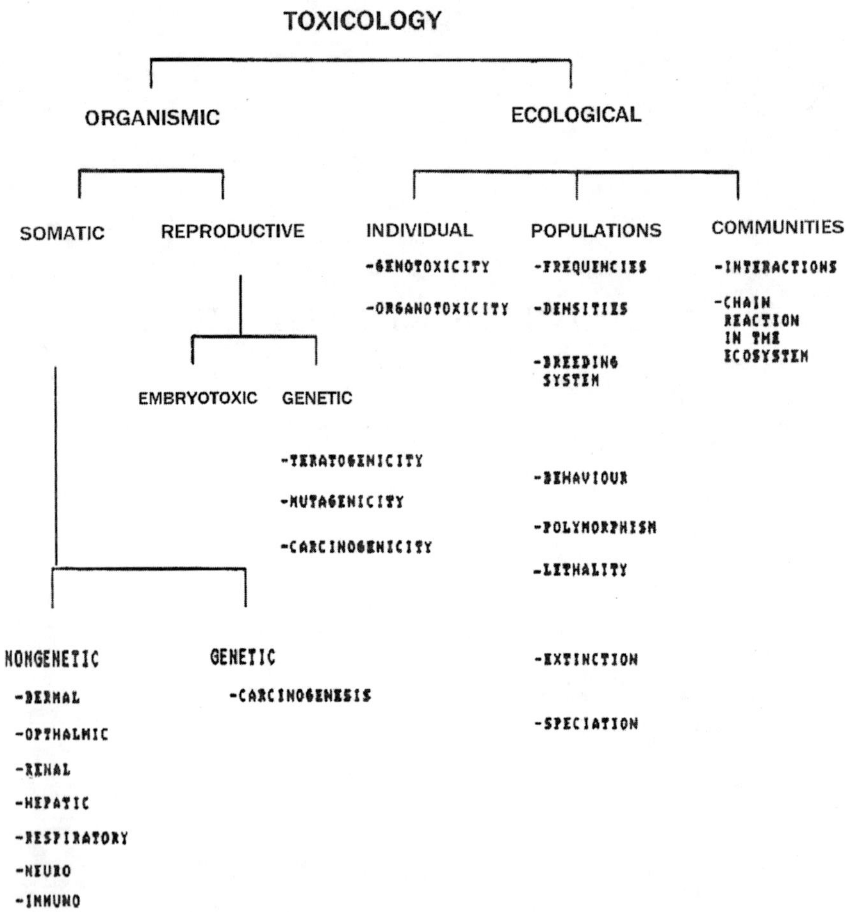

DOCUMENT - 2

GENOTOXICITY IN THE ECOSYSTEM : THE COMPONENTS

- HABITATS
  - Air
    - Ambient
    - Occupational
  - Water
    - Pond
    - Lake
    - Estuary
    - River
    - Sea
  - Agroecosystems
  - Waste lands
- SOURCES
  - Pollution
    - Air
    - Water
    - Soil
  - Agriculture
    - Pesticides
    - Sewage
    - Effluents
  - Metabolic activation
    - Denizen biota
- PURPOSES
  - Low dose effects
  - Food chain risks
    - Plants
    - Animals
  - Epidemiology of disease
  - Impacts on Ecology
  - Evolutionary variations
  - Management
  - Preventive medicine
- SYSTEMS
  - Plants
  - Animals
  - Humans
  - Microbes
- APPROACHES
  - Denizen
  - Sentinel
  - Habitat sampling
  - Concentrator Assay coupling

DOCUMENT - 3

GENOTOXICITY IN THE ECOSYSTEM

THE MONITORING APPROACHES

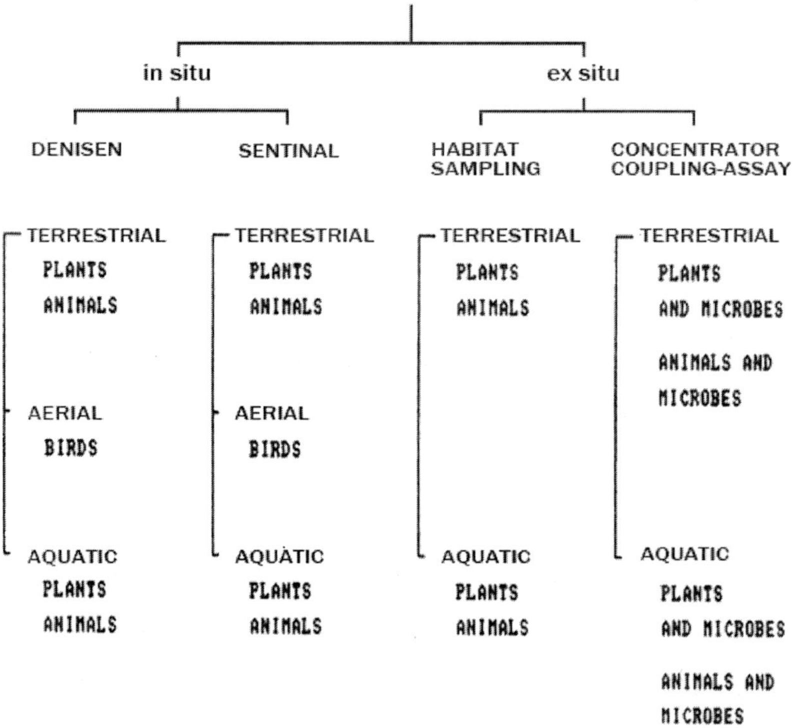

DOCUMENT - 4
MONITORING SYSTEMS

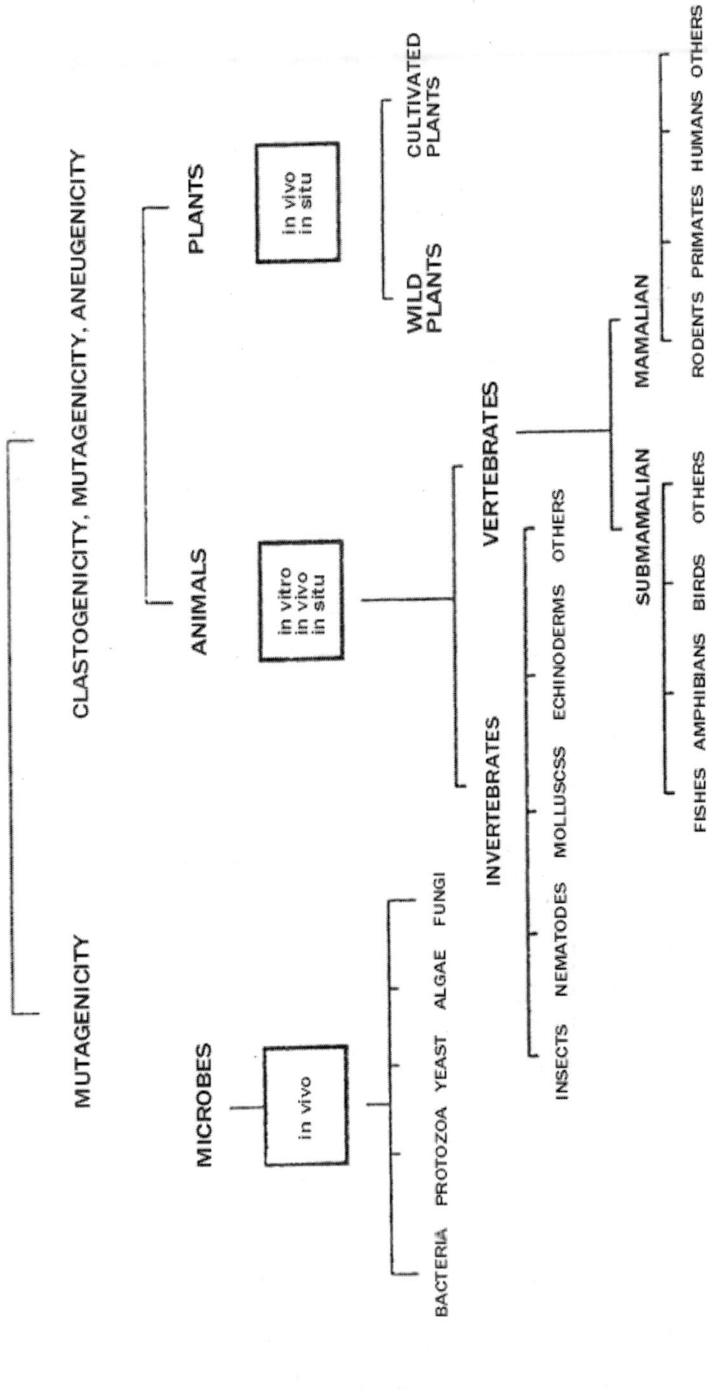

DOCUMENT - 5

CELLULAR ENDPOINTS INDICATING GENOTOXIC RESPONSES TO STRESSES BY THE MONITORING SYSTEMS

SCORABLE ENDPOINTS

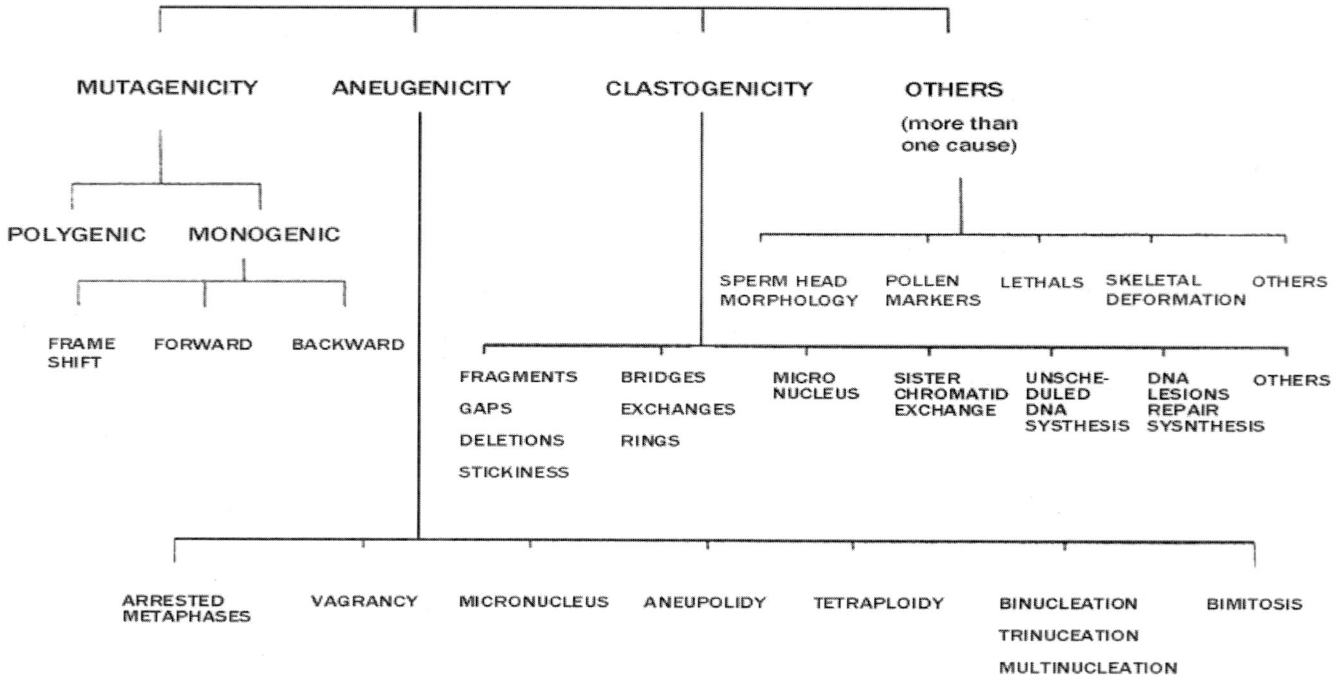

DOCUMENT - 6

BIRTH DEFECT SYNDROMES INDICATING GENOTOXICITY

Genomic	**Down, Patau, Edwards, Kilnefelter, Turner**
Genic	Lou-Gehrig, Tay-Sach, Pentosuria, Polycystosis, Retinitis pigmentosa, Multiple exostoses, Spinocerebellar ataxia, Cystic fibrosis, Muscular dystrophy, Ataxia, Xeroderma pigmentosa, Sickle Cell anemia, Huntington's disease, Alkaptonuria, Hemophilia, Phenylketonuria, Galactosemia, Thalessemia
Chromosomal	Bloom, Buckling, Cri-du-chat, Philadelphia
Others	Dwarfism, Cleft lip and palate, Sperm abnormalities, Limb and skeletal deformities
Total Syndromes	5000
Frequency of each syndrome	1/100,000 to 1/1000
Total Cancers	>200 Frequency: 2.5 to 5% in the west Monitoring
	Birth or population Frequency: 10-15%

DOCUMENT - 7
SOME GENOTOXIC CHEMICALS

TYPES	EXAMPLES
1. Mutagens	Formaldehyde, Chloramphenicol, Hydroxylamine, Styrene, Benz(a)Pyrene, Sodium Azide, Ethylenimine, Methane Sulfonates
2. Clastogens	Mitomycin C, Actinomycin D, Isoniazid, Benomyl, Chloramphenicol, Ethidium bromide
3. Aneugens	Ethylene glycol, Griseofulvin, Nitrous oxide, Lindane, Chloral hydrate, Ethoxy Caffeine, Ethidium bromide
4. Promutagens	Cycasin, Saffrole, Vinyl Chloride, Morpholins, Olifins, Piperidines
5. Antimutagens	Vitamin B, Vitamin C, B-Carotene, Theophylline, Rotenoids, Biometals
6. Supermutagens	Sodium Azide, Ethylenimine, Methane Sulfonates
7. Carcinogens	Saffrole, Sodium Azide, Benz(a)Pyrene, Cyclamates, Furfuryl Furamide, Arsenic, Benzene, Chromium, Nickel, Aflatoxins, Nitrosamines, Epoxides
8. Teratogens	Thalidomide, Hycanthone, Methyl Mercury, Tetracycline, Chloroquin, Mimosine, Bisulfan, Nicotene, Caffeine

DOCUMENT - 8
SOURCES OF GENOTOXINS IN THE HUMAN ENVIRONMENT

General Types	Specific Classes
1. Naturals	Alkaloids, Glycosides, Flavonoids, Others
2. Foods	Colours, Sweeteners, Sterilants, Additives, Processing
3. Habits	Mastications, Inhalations, consumptions
4. Pesticides	Insecticides, Nematocides, Fungicides, Herbicides, Acaricides
5. Drugs	Modern, Traditional
6. Cosmetics	Natural, Synthetic
7. Dyes	Consumptives, Nonconsumptives
8. Industrials	Pharmaceuticals, Rubber, Textiles, Petroleum, Paper fertilizers, cement, explosives, Leather, Photography, Glass & ceramics, resins, sugar, Paints, Metallurgy, Iron & Steel, Plastics, others
9. Energy	Nuclear, Petroleum, Biogas, Biomass (Production, Conversion, usage, Transport)
10. In Situ Radiation & Chemicals	
11.Occupational Heavy metals, Aromatic Hydrocarbons, Halogenated & Aromatics, Nitroso Compounds, Combustion	Products & Ambient Air
12. Water	Domestic and Industrial Effluents
13. Soil	Pesticides, effluents, Others
14. Food Chain	Residues of Toxic Pollutants
15. Metabolites	Synergistics, Degradations, Activations
16. Others	

DOCUMENT - 9
SOME GENOTOXIC STRESSES TO INDIAN POPULATIONS

Agents	Occurrence	Exposure	Effects
1. B&G rays	monazite sands of kerala beaches	thousands of people for variable durations	birth defects detected
2. X-rays	50 000 medical facilities	a few lakhs and for many years	impact reports under scrutiny by AERB
3. UV-rays	Fall out in southern most India	millions and for variable durations	spurt in skin cancers feared
4. EM-rays	electro magnetic appliances at homesteads	millions and for variable durations	effects suspected
5. Flavanoids etc	natural in vegetables	millions of consumers and life long - convertible into genotoxins	effects suspected
6. Alkaloids	Tobacco-areca nut-betel leaf quids	millions of habituals for brief spells	conform association with cancers in a few thousands of people, more suspected
7. Aflatoxins	fungal toxins as contaminants in cereal grains	Uncertain size of consumers for limited durations	confirmed carcinogens
8. Pesticides	contaminant residues in the vegetable foods	millions of consumers for long spells many confirmed genotoxins	birth defects reported
9. Nitrosamines	variety of vegetable and animal foods, beers, wines and liquors	millions of indiscrete consumers	confirmed genotoxins, victims suspected.
10. Nitropyrenes	reprography, diesel exhaust, combustion of biomass, tobacco smoke	millions of occupational workers	BaP is a confirmed genotoxin, victims suspected.
11. Many dyes, Nitrates, AF2 etc	in fast and canned foods, soft drinks, ice-creams and others as colourants, preservatives, flavonoids, etc	a few millions of modern gastronomical consumers	many are confirmed genotoxins. victims suspected

(Contd...)

12. Deformed amino acids, pyrenes and anthracenes, aromatic hydrocarbons	deep fried and pyrolyzed foods	millions of consumers and cooks	confirmed genotoxins, victims suspected
13. Chlorinated Compounds	Purified municipal water supplies	All direct consumers	some are confirmed genotoxins, victims suspected
14. Drugs	therapies against infections (malaria, helminth, tubercle etc) cancer	a few million patients for brief spells	some drugs are confirmed genotoxins, victims suspected
15. Heavy metals	Occupational environment of many industries, cosmetics, diesel and cigarette smokes	millions of workers in many industries, cosmetic users, smokers etc	confirmed genotoxins, victims suspected
16. Styrene and SO2	Occupational atmosphere of many industries	thousands of workers for specific durations	confirmed genotoxins, victims suspected
17. Arsenic	ground waters of Bengal	millions of consumers	effects expected
18. Others	40 atomic energy establishments with possibilities of radiation exposures; nine coastal zones with polluted aquatic food chains; 16 urban centers where most genotoxicity may occur; vegetarian diets contain genotoxins as well as antigenotoxins; Indians are after the same genotoxic practices which the westerners try to abandon.		

DOCUMENT - 10

- in situ Investigations: A summary of Animal studies

Tests	Habitat	End point	Result	Reference
1. Anodonta cygnea	Waste water Chlorination, Italy	Micronuclei	+	Scarpato et al 1990
2. Bufo melanostictus	Polluted ponds, India	Sister Chromatix exchange	+	Chakraborti & Banerjee, 1984
3. Chironomus tentans	Radioactive wastes in a lake, U.S.A.	Polymorphism, Invesrsions	+	Blaylock 1966
4. Chrotogonus saussurei	sewaged habitats India	Meiotic abeberations	+	Sharma et al 1988, 1989
5. Clupea harengus	Surface micro layer, Baltic	Embryo deformities	+	Landolt & Kocan 1985
6. Dipodomys merriami	Military smoke, USA	Sister Chromatix exchange	+	Sandhu & Lower 1987
7. Dipodomys deserti	Military smoke, USA	Sister Chromatix exchange	+	Sandhu & Lower 1987
8. Gadus morhua	Surface micro layer, Baltic sea, harbour, Germany	Embryo deformities	+	Landolt & Kocan 1985
9. Littorina saxalities	Domestic and industrial wastes, UK	Embryo deformities	+	Dixon & Pollard 1985
10. Mus musculus	Polluted air, Ontario	Embryo deformities	+	Nayak & Petras, 1985
11. Mytilus edulis	Harbours, UK	Aneuploidy	+	Dixon 1982
12. Mytilus galloprovincialis	Polluted adriatic sea, Yugoslavia	Chromosomal aberration	+	Al-sabti & Kurelec, 1985
13. Mytilus galloprovincialis	Seaports, Italy	Micronuclei	+	Scarpato et al 1990
14. Nephthys incisa	Harbour sediment	Sister chromatid exchange	+	Pesch et al 1987
15. Perca fluvialitis	Paper mill effluents Baltic Sea, Sweden	Micronuclei	+	Al-Sabti & Hardig 1990

(Contd...)

16. Peromyscus leucopus	Waste land -fills, USA	Micronuclei Sister Chromatid exchange	+	Tice et al 1987
17. Poecilocerus pictus	Urban pollution, India	Meiotic aberrations	+	Shyama & Rahiman, 1987 a,b
18. Psettichthya melanostictus	Surface micro layer in Harbour USA	Chromosomal Aberration	+	Landolt & Kocan, 1985
19. Salmo gardneri	Surface micro layer in Harbour USA	Chromosomal Aberration	+	Landolt & Kocan, 1985
20. Salmo	River sediment, Buffalo, USA	Embryo deformities	+	Maccubin et al, 1985
21. Sigmadon hispidus	Hazardous wastes, Texas, USA	Chromatid aberrations	+	Thompson et al, 1986
22. Umbra limi	Polluted Rhine water	Sister Chromatid exchange	+	Alink et al, 1980
23. Demersal* Fishes	Urban wastes Canada, USA	Micronuclei	+	Cross & Hose, 1986

DOCUMENT - 11

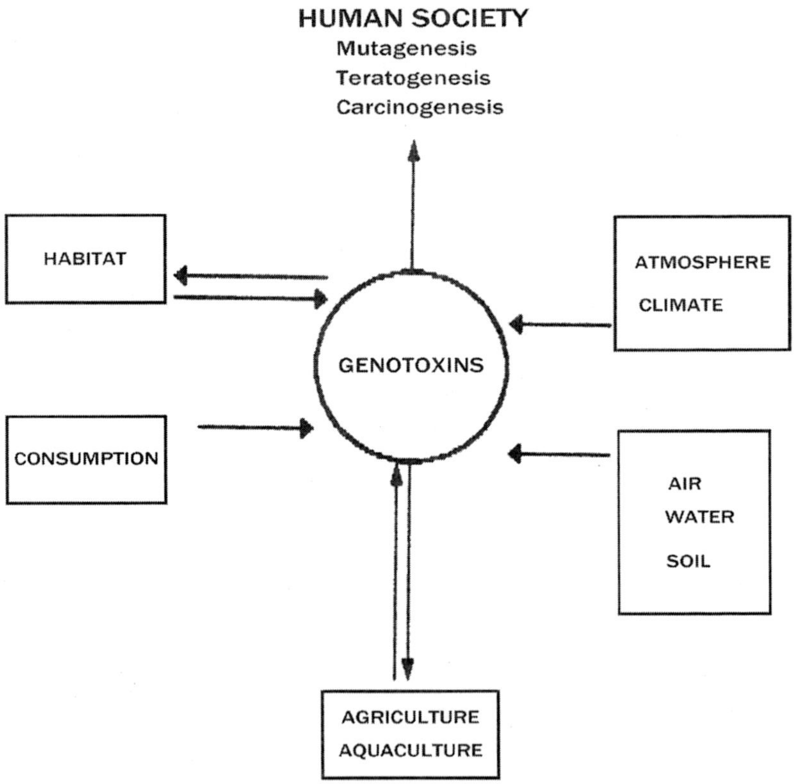
HUMAN SOCIETY
Mutagenesis
Teratogenesis
Carcinogenesis
HABITAT
ATMOSPHERE
CLIMATE
GENOTOXINS
CONSUMPTION
AIR
WATER
SOIL
AGRICULTURE
AQUACULTURE

DOCUMENT - 12

SUSCEPTIBILITY OF HUMAN FEMALE TO MUTATIONAL INSULTS FROM CHEMICALS IN THE ENVIRONMENT

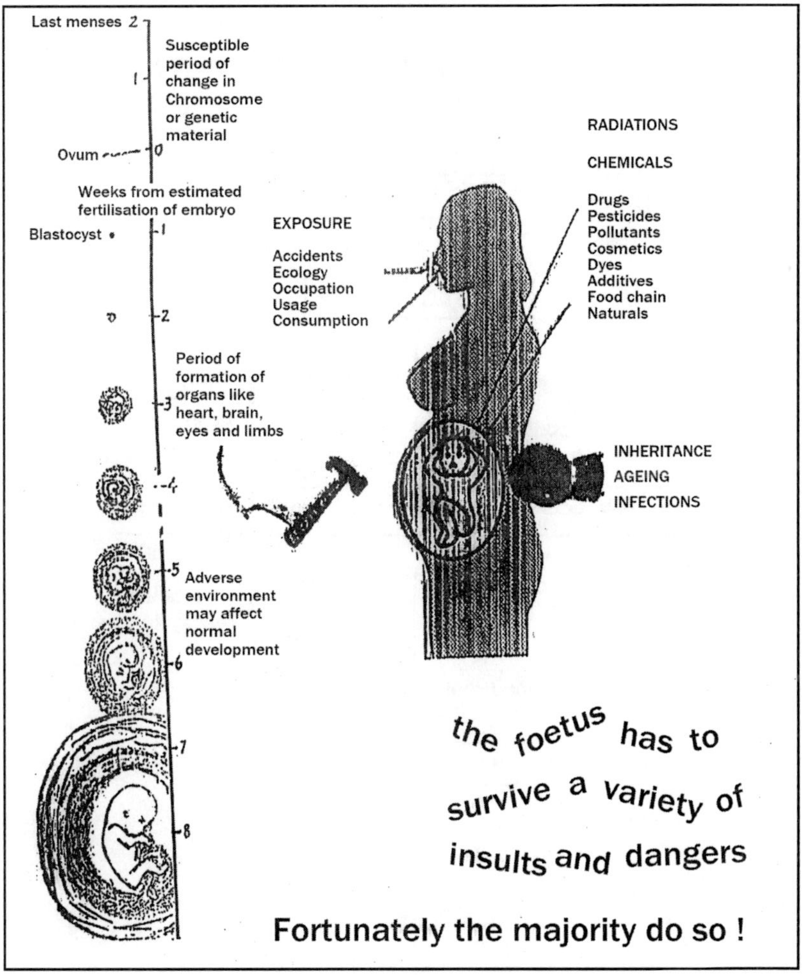

DOCUMENT - 13

INSTABILITY IN THE AGROECOSYSTEMS DUE TO PESTICIDES

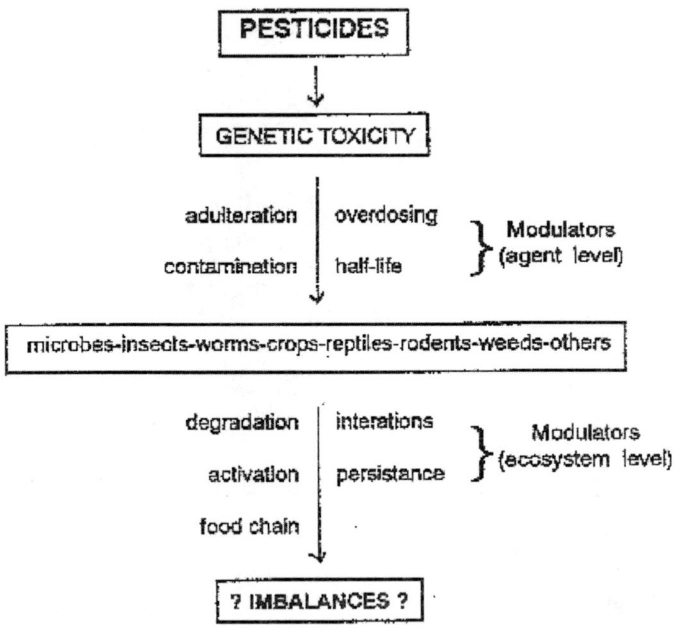

DOCUMENT - 14

THE FIRST CENTRAL DOGMA OF GENETIC TOXICOLOGY

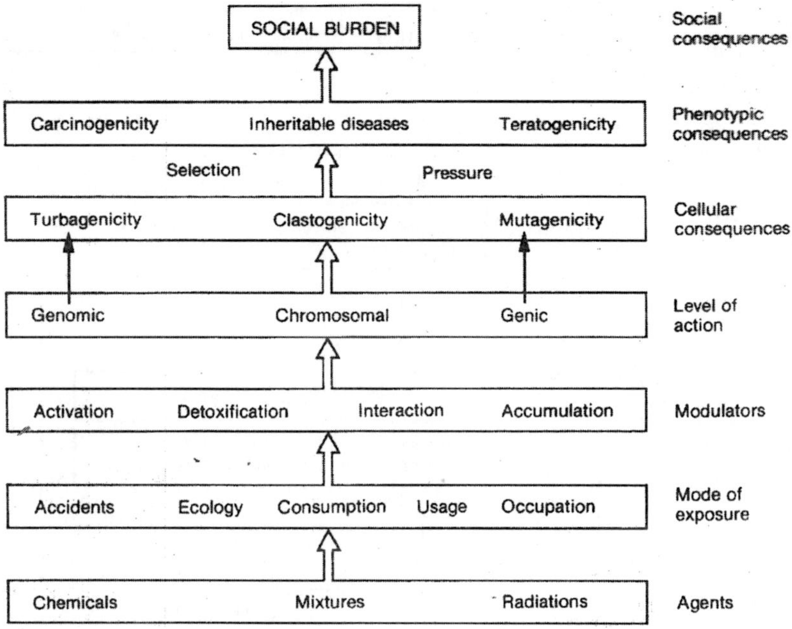

DOCUMENT - 15

THE SECOND CENTRAL DOGMA OF ENVIRONMENTAL MUTAGENESIS [2000 AD]

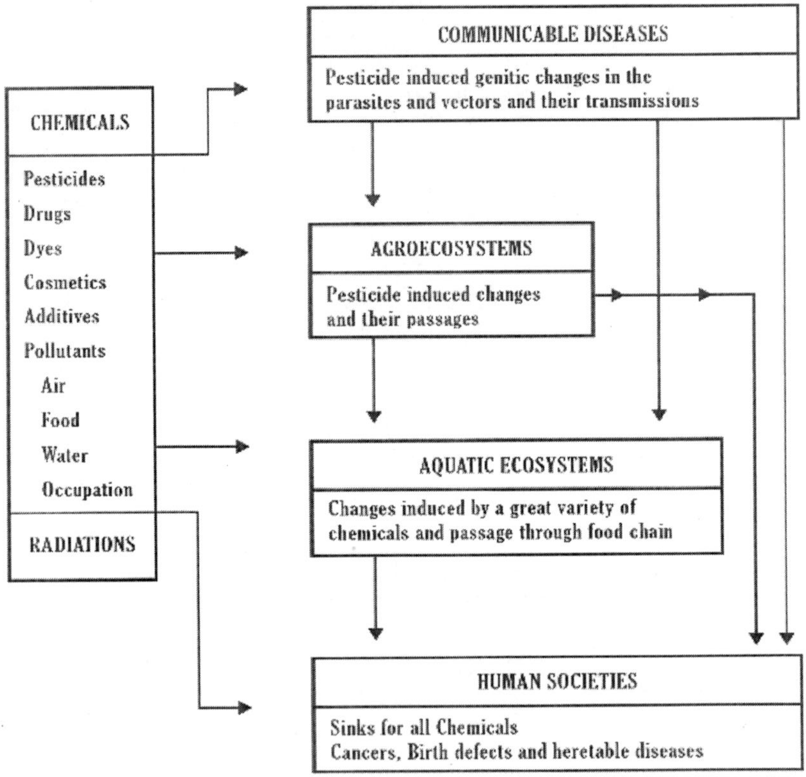

DOCUMENT - 16

GENOTOXIN DYNAMICS IN THE ECOSYSTEMS ORIGINS (1), FLOWS (2), POOLS (3), SINKS (4) PROBES (5) EFFECTS, AND (6) HUMAN MISERIES

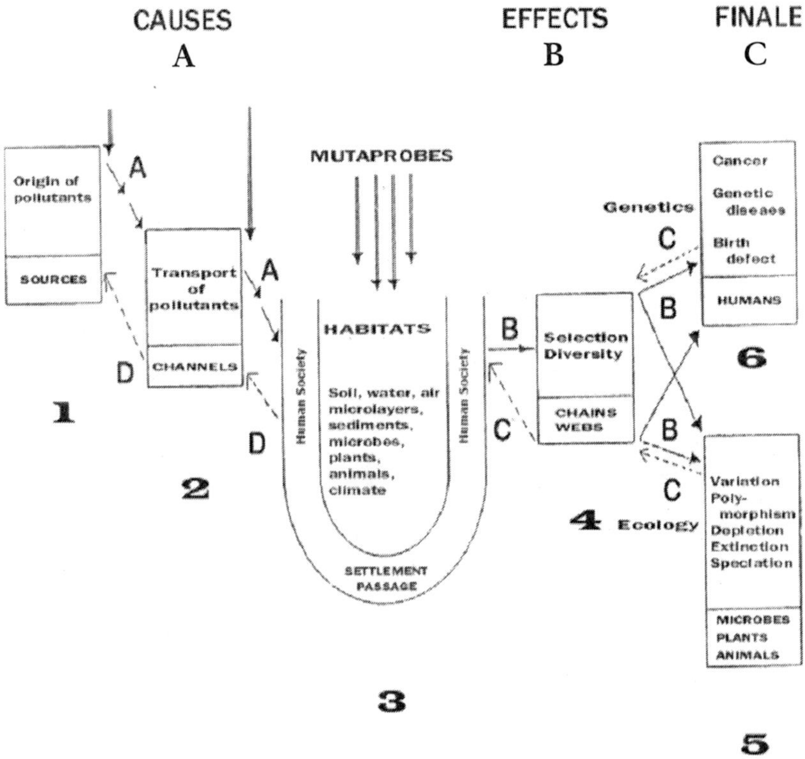

DOCUMENT - 17

AQUATIC HABITATS OF INDIA WITH GENOTOXIC POTENTIAL

Coastal Regions

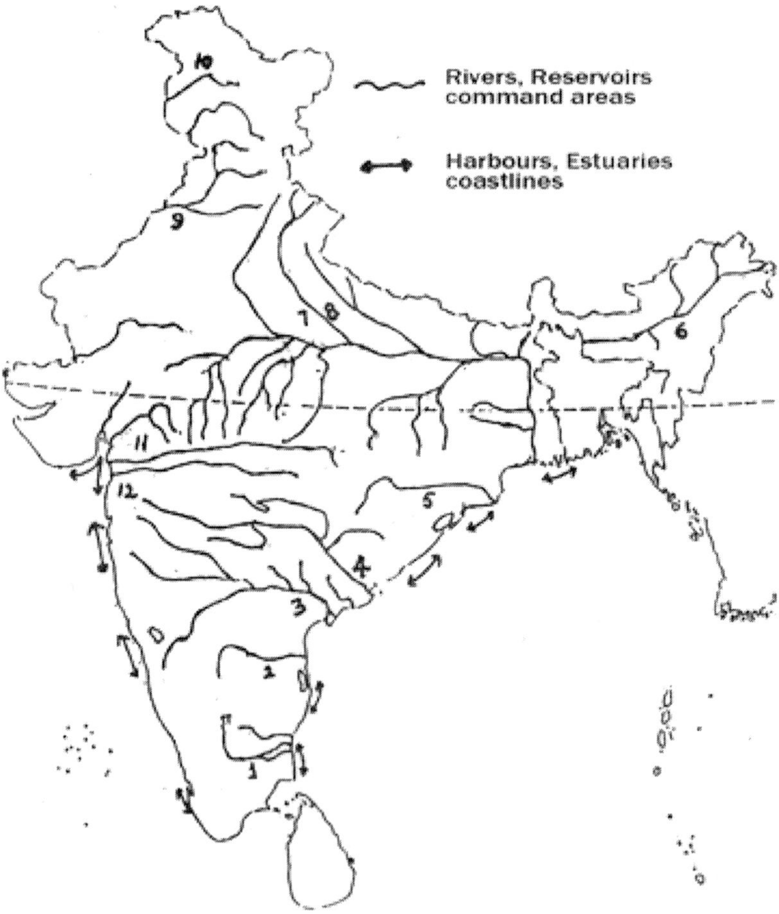

DOCUMENT - 18

AQUATIC HABITATS OF INDIA WITH GENOTOXIC POTENTIAL

Riverine Sites

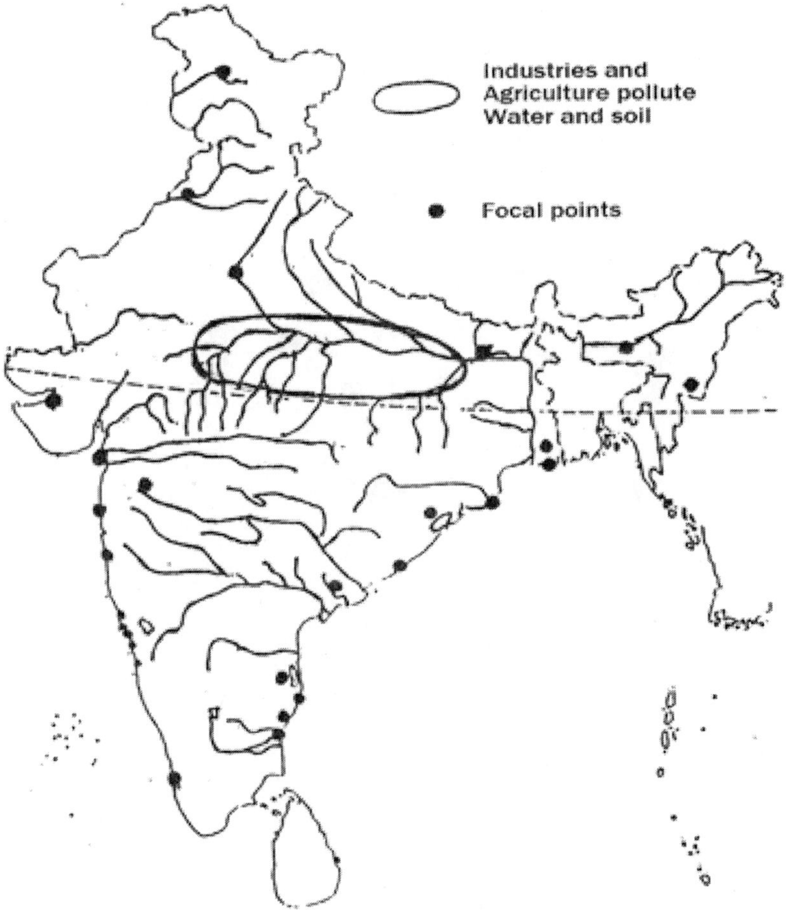

DOCUMENT - 19

URBAN POLLUTED HABITATS OF INDIA WITH GENOTOXIC POTENTIAL

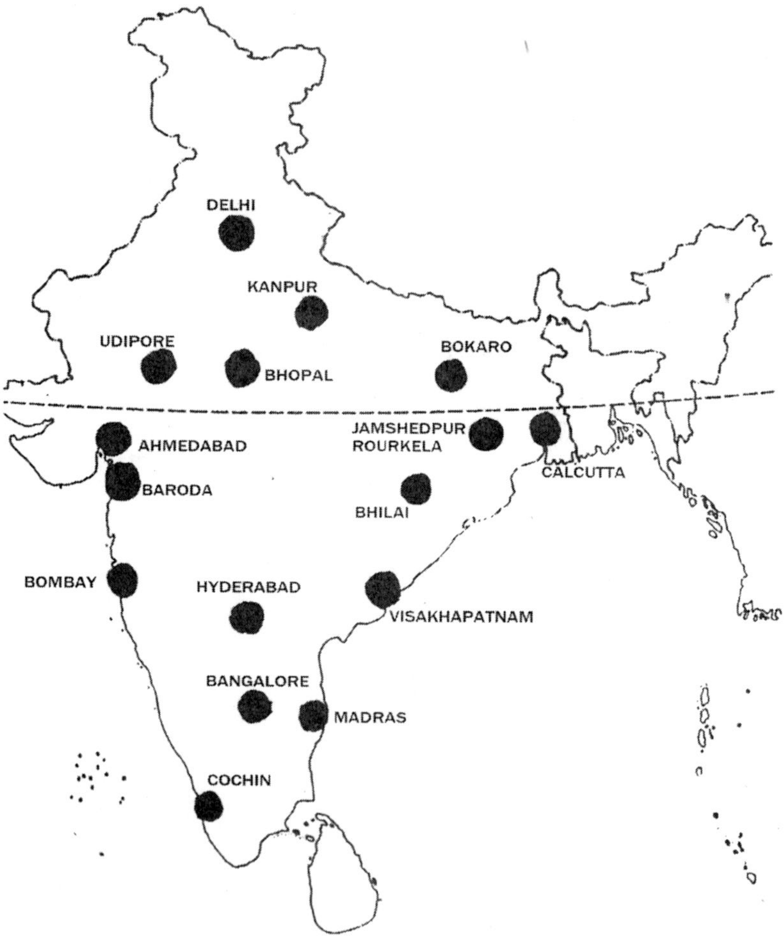

THE IMPORTS OF THESE DOCUMENTS CAN BE SUMMARISED AS FOLLOWS

1. By 1969 a Society on Genetic Toxins was formed in USA, called EMS, because Ethyl Methane Sulphonate [EMS] was the most popular and potent mutagen in research and applications at that time. Otherwise it was more a Genetic Toxicology than Environmental Mutagenesis. This was the time when the world was witnessing disastrous events in its environment. By the time EMSI was initiated in 1975, basic concepts were being agreed to yield a conceptual framework. Chemicals in direct use by humans were the focus – the life styles. This became the central dogma.
2. Almost simultaneously and thereafter, focuses were centering on Pesticides, which are designed to be toxic. The primacy was shifted from life styles to Agro-ecosystems. This is for 2 reasons. Crop yields on which human lives depend may be reduced due to undesirable resistances to pests and sensitivities of crops, genetically. Further such pesticides could pass to human systems through food chain becoming a life-style genotoxin - as residues or metabolites.
3. Pesticides would mean not merely residues but also their degraded or modified substances. Food chains need not be short but substantially long, thus paving the way to any other substances in the related systems. Thus the agents, routes and impacts become complex. All these force revising the earlier premises and other related aspects consequently. And more systems and pollutants became relevant to genetic toxicology. The subject now becomes truly Environmental

Mutagenesis! Thus the theory of Gene-Epidemics had to make an entry !

4. *In situ* studies further changed the scenario to greater complexity but nearer to the total truth. Economies becoming capitalistic, developing countries turning globalized, production, consumption, buy and sell syndrome becoming our new Gods during the last 3 decades. Consequently the number of chemicals released into air, water and food stuffs became multifold. They could genetically affect not only humans, but any organism – small or big – resulting in a swarm of mutants, These could impact any other organism – importantly, the humans. This is unforeseen but unpreventable – bolts from the above. The present Covid is one like this - a Mutant Virus about which most did not know anything until human health is impaired. Similar ones may be in the waiting to appear in due course. – the Coming Plagues, Laurie Garret termed!
5. Thus it would appear that Environmental Mutagenesis has greatly expanded in 50 years – first limiting to Human Societies, then Agrosystems, and appeared to end in Polluted sites. But now it appears that the spread is all-encompassing all environments, all food chains, and all organisms; further some of these organisms damaging the humans- truly epidemic and pandemic ultimately. A whole ominous subject is awaiting for detailed studies with an Ecosystem Approach, for, the Gene- Epidemics have that much expanse and depth to probe.

DOCUMENT - 20

GENE-EPIDEMICS : THE CURRENT CENTRAL DOGMA

Sl. No.	Systems	Exposures	Ultimate Effects
1.	Life Styles	Foods Food Chains Usages Pollutions Metabolites	1. Detriments 2. Cancers 3. Birth Defects 4. Inherited Diseases
2.	Agriculture Aquaculture Veterinary Practices Other Cultures	Pesticides Metabolites Interactions	5. Resistance of Insects, fungi, Weeds - Metabolites 6. Sensitivity of useful Microbes Metabolites 7. Sensitivity of high yielding crop varieties
3.	Health Care Systems	Drugs Metabolites	8. Pathogens' Resistance 9. Parasites' Resistance 10. Induction of New systems
4.	Public Health, Social and Preventive Medicine	All the above	11. New strains of Viruses and Bacteria such as Escherichia, Salmonella, Pneumococcus, Influenza, Corona, etc.
5.	Ecosystems	Any Chemical on any organism with any effect towards any consequence to the system	12. Biodiversity 13. Mutation Pollution causing imperceptible effects on Human Society

6. **End of Civilization May begin with Gene-Epidemics.** There is a considerable body of opinion that the beginning of our civilization has begun. This kind of future was envisaged by some of the votaries of LTG world view of COR advanced as back as 1972. But the same authors, years later prophesied that this can be managed with sufficient efforts in several human activities. This was criticized for its Malthusian content. Interestingly the doomists predicted the collapse of civilization by the middle of this century, starting from the third decade ! The issue is about the symptoms.

Climate Change is prominent in this scenario. Forest fires, Rains and Floods, Crop failures, Human Migrations – in short all disasters, But I consider the damages from the Microbes as the most telling. Garrett was suggestive of this in 1996, and Covid is here. I envisaged this possibility in 1985 in agro-ecosystems, extended it in 2002 to ecosystems and developed the theme by 2009 when we predicted Gene - based epidemics which could become Pandemics.

Basis for these are data and reasoning about possibilities. In fact the situation about mutations is that many chemicals in small doses and complex mixtures pervading the ecosystems can cause mutations in any organism through a vcariety of exposures.

To a positive mind all problems are surmountable by human intelligence and endeavors but for the politics among the rulers and among the ruled. The former is afflicted by immoralities currently experienced prominently in the state of AP, soon to spread out. The latter pervaded through out India during globalization of 30 years and recently and more vigorously since march this year. When psyches are altered, cultures vanish, political suicides become unpreventable. This book in three parts, is meant to vent out my forebodings. Therefore it may be personal, anecdotal but scientific as well as experiential. Hope there will be some souls to share my pains and views, provided providence grants me the time to complete this venture!

B. GENERATORS OF KNOWLEDGE

Scientist in this field assembled periodically at Bombay, Budapest, Edinburg, Penang, Bangkok, and Cleveland

The first meeting of the Environmental Mutagen Society of India formed at Baroda University. The first office bearers of the Society were: Gopal Ayengr, Sundaram, Notani, Aiyar, Bhisey etc. Most of the members came from BARC, Cancer Research Institute at Bombay, and other allied Bodies from other places.

Sharma with Prof. Gyula Ficsor of Kansas University, who was basically a Botanist from Hungary, Prof Malling, an originator of EMSUSA is in the background.

Prof. Tazima of Japan famous for Silkworm Genetics with Prof. Charlotte Auerbach who first discovered that chemicals can induce Mutations in Dosophila in Edinburgh University during 1940s.

Prof. Alexander Dubinin, Director of institute of Genetics, Moscow with Alexander Hollaender of Oak Ridge National Laboratory. The former had to spend in gaol fo years during Stalinist regime for believing in Mendelian Genetics!

Dr. Bridges, a Microbiologist and Prof. Kilbey, well known for his treatise on Genetic Toxicology Test Procedures.

Prof. Samuel Epstein the first Secretary to the Environmental Mutagen Society, USA, with some younger scientists, such as Castleman, at the dining space.

Dr. Kaja Linnainmaa and Sharma, two developing scientists from Finland and India with a senior scientist, Prof. William Grant of McGill University in Quebec, Canada.

Dr. Rosalie Bertrell has been a Radiation Biologist of Canada, turned a Campaigner against ionizing radiations, In the later years.

Dr. Lalji Singh in the Institute of Genetics, Edinburgh University, who created the phrase "Snake's Eye View" in Nature. Sharma was his guest at Edinburgh briefly. Lalji became Director of CCMB, Hyderabad and VC of BHU, Varanasi, subsequently. I had then a very constructive and successful role in one of his major projects, appreciated by all friends and leaders at that time!

Queing for Dinner. Prof. Dubinin, Prof. Borstel of Canada in the background Dr. Marza Sorsa from Finland. in whites.

Prof. W. F. Grant of McGill University, Canada with Prof. Tazima of Japan, and Dr. Hollaender of USA, with their spouses.

Budapest University. Dr. Sharma spent some valuable time with Prof. Vida Gabor And his team of workers. Dr. Siddique of Patna was one of them near the bottom.

Near the Dance floor at Edinburgh. Most youth were here and enthusiastic watching the Scottish National Troup Performing

Sharma with two Botanists from Poland and Sweden. The latter Dr. Gerid Fiskesjoe, was a student of Prof. Albert Levan continued his tradition

In an outdoor cultural Programme are seen Prof. A T Natarajan from Leiden in Netherlands, A colleague of him from Leiden, another scientist from Egypt and Dr. Gopalan from Nairobi.

Sharma and Dr. Prakhya B.K. Murty in Cleveland near to the Lake Eerie waterfront . He was the director of Frederick Institute of Toxicology near Tambaram mostly focusing on pesticides.

Dr. C.R Krishnamoorthy was the director of CTRI., Lucknow. He was the leader of the investigation on Bhopal Gas Tragedy. He has initiated many studies in aid of BGT.

Sharma with Prof. Y R Ahuja on the shores of lake Eerie at Cleveland. Prof. Ahuja worked at Osmania University mostly on human *in vitro* systems.

An Egyptian scientist with Dr. Mendelsohn one of the organizers of the conference and Dr. Brusick. Dr. Brusick wrote a very timely and useful book on Genetic Toxicology very handy and useful to all.

Dr. De Serres and Prof. Sobels of Netherlands. Dr. De Serres is the co-editor of Chemical Mutagens in 10 volumes along with Hollander. Sobels was the Chief Editor of the journal Mutation Research.

The fourth International conference was inaugurated by the President Dr. Frederick de Serres in a Chapel situated with in the campus of the University of Cleveland.

Conference Hall with Dr. James Gentile in the front and Dr. Sankarnarayanan to his right, above. The former has well known for his studies involving plants along with Dr. Plewa. The other, originally from India has been working in Netherlands. He was famous for the concept of Detriment Index developed along with Dr. Czizel of Hungary.

Many Indians can be seen interacting with each other - some new to USA and others.

Prof. Ehling of Germany with his wife At the background are Prof. Rosenkranz and Prof. Claes Ramel of Sweden, in serious discussions.

Dr. Bruce N Ames is the most famous of all Scientists in Environmental Mutagenesis. This is for the test he developed for quick identification of Mutagens. And it was proved to be right all over the world.

Mixing food and Business during the ICEM are an Indian Scientist, Prof. H K. Goswami of Bhopal with Dr. Sandhu making a name in the USA. He was originally from the Amritsar School of Botany and Genetics.

Enjoying a Dining together Dr. P. C. Chauhan of India, Dr. Rosenkranz and Mendelsohn of USA. Rosen kranz working at Cleveland University, was famous for studying structure - activity relations and using computer systems for this and allied works.

Dr. Sharma of India was showing to the photographer the serious discussions between Profs. Rosenkranz and Mendelsohn, the chief organizers of the conference.

These are the triumvirate of the conference - Rosenkranz and Mendlesohn sitting and Federik De Serres standing all watching intently.

In the Conference Hall are seen many Scientists. Dr. De Serres, the President of Conference and Dr. von Borstel from Canada, can be seen in the first row.

A Ceremonial dining table accommodating Dr. Kada of Japan, Drs. Notani and Sharma of India, Dr. Tazima of Japan, Drs. Rosenkranz and Mendelsohn of USA and Prof. Ehling of Germany.

Prof. Rosenkranz sitting, while Prof. Sobels fom Netherlands and Dr. De Serres appear to be sharing the moments. Prof. Sobels is a leader in Europe on mutagen studies and a world leader in Mutation Research Publications.

One of the quietest persons at the Mutagen Conference was Prof. Chang of China standing along an Egyptian Scientist near the Lake Eerie in Cleveland.

Stalwarts on their own to whom I waqs happy to get interested in the genetic effects of pesticides and other chemicals. They are Dr. Vandana Shiva, Mohammed Idris of Malaysia Sunderlal Baguguna, and Nagarajan.

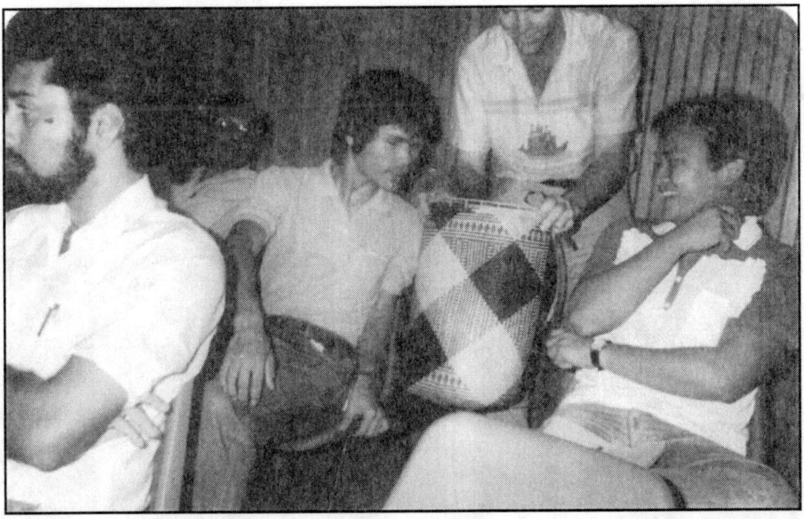

At the centre is Mr. Harrison Ngao, a local tribal of Sarawak, an Indonesian part od Malaysia. He was recognized for his unusual gifts of leqdership and awarded Magsaysay Award for successfully resisting exploitation of his people. I met him at Penang along with Mohd. Idris, and Martin Thong.

At Penang were several anti-pesticide campaigners who became interested in genotoxicity of pesticides. They are Solomon, Patricia, Gerrits, Hontelez, Eloy and Palleemerts, from left to right.

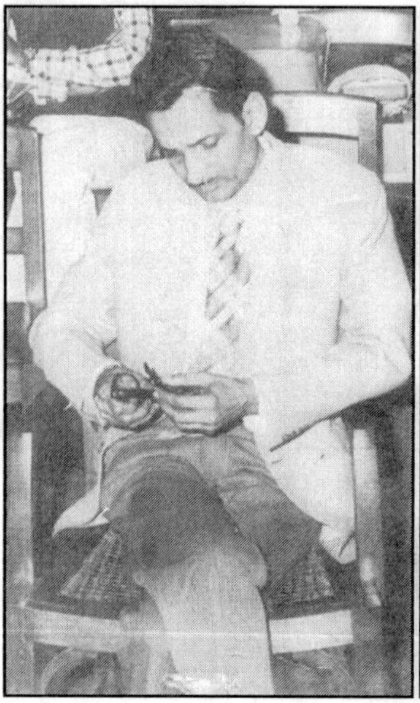

A.K. Sharma

M.S. Swaminathan

M.V. Murty

N.K. Notani

Gopal - Ayengar

K. Sundaram

Lalji Singh
C.B.S.R. Sharma

Nagarajan

A T Natarajan

Sankar Narayanan

B.N. Ames

Brusick

Fred De Serres

A. Hollaender

Ehling

Sobels

APPENDICES

1. CORONA, COVID AND MUTATIONS
2. PRICE OF SILENT MUTATIONS
3. LAURIE GARRETT
4. U.S.-CHINESE MUTUAL DISTRUST ON THE ORIGIN OF COVID

1
A PLENARY LECTURE

By Prof. CBSR. SHARMA

45TH INDIAN SOCIAL SCIENCE CONGRESS

FOCAL THEME

ENVIRONMENT, ENERGY AND HEALTH

(B.S. Abdur Rahman Crescent Institute Of Science & Technology, Vandalur, Chennai 600048)

MARCH 28 - APRIL 01, 2022

A PLENARY LECTURE

COVID REVEALED CORONA AND MUTATIONS

CORONA AND MUTATIONS UNFOLDED GENEPIDEMICS

GENEPIDEMICS THREATEN AND FRIGHTEN FUTURES

1. CORONA, COVID AND MUTATIONS

INTRODUCTION

When India gained independence from the British and formed its own policies and administrations, despite good hopes and best wishes there appeared three lacunae and continued forever. These are the systems of Health, Education and Justice. They continued to deteriorate and everyone reconciled and ultimately acquiesced with realities.

The second freedom was economic 45 years later in the form of LPG. Since a free for all becomes a state policy, governments had to take certain precautionary decisions in the interests of people's welfare. For example in the fields of education and health there ought to be intensive development in sensitive areas. These are about" things we cannot see and sounds we cannot hear" as Russians used to call., because these could enter our ecosystems from alien or foreign lands by chance or design and play havoc with people. This happened in 2019, in the form of COVID-19 that does not seem to end but only pause.

COVID-19 is the impact of a Variety of a virus called Corona. This variety was soon discovered to be a mutant of the wild type, through a process called MUTATION, well established in Biology. But mutations had to be induced or produced. If induced in nature it is called spontaneous mutagenesis through chemicals or radiations. If produced it ought to be planned and executed by the ingenuity of human mind.

Based on its behaviour and progression, Covid-19 mutant appears to have been consciously developed in a laboratory than originated spontaneously.

If the first was taken care of, the second would not have been required, if the second was attended to, the third would not have been so severe and pervasive. The responsibility for the first was of political leaderships, and for the second, both the Political and scientific leaderships.

And for the last, all are accountable!

At present I wish dive into this third aspect

It begins with Corona, Moves into Covid, Reveals Mutations and renders Futures OMINOUS THROUGH GENEPIDEMICS Corona viruses are a group of RNA viruses that cause respiratory tract infections ranging from mild cold to lethality. Lethal varieties can cause SARS, MERS and COVID-19, which is causing an ongoing pandemic.

They are enveloped viruses with a single-stranded RNA genome and a nucleocapsid of helical symmetry. The genome size ranges from 26 to 32 kilobases, one of the largest among RNA viruses. They have club-shaped spikes that project from their surface, which in electron micrographs create an image reminiscent of the solar corona.

The name "coronavirus" is derived from Latin *corona*, meaning "crown" or "wreath", itself a borrowing from Greek κορώνη karōnē, "garland, wreath". The name was coined by Almeida and Tyrrell who first observed and studied human corona viruses. The name refers to the characteristic appearance of virions by electron microscopy, which have a fringe of large, bulbous surface projections creating an image reminiscent of the solar corona or halo.

As of 2020, 45 species are officially recognised.

HISTORY

The earliest reports of a coronavirus infection in animals occurred in the late 1920s, in North America. Human coronaviruses were discovered in the 1960s. They isolated and grew the virus in kidney tissue culture, designating it 229E. The novel virus caused a cold in volunteers and was inactivated by ether.

Human coronavirus 229E and human coronavirus OC43 continued to be studied in subsequent decades. The coronavirus strain B814 was lost. It is not known which present human coronavirus it was. Other human coronaviruses have since been identified, including SARS-CoV in 2003, HCoV NL63 in 2003, HCoV HKU1 in 2004, MERS-CoV in 2013, and SARS-CoV-2 in 2019. There have also been a large number of animal coronaviruses identified since the 1960s.

MICROBIOLOGY

Coronaviruses are large, roughly spherical particles with unique surface projections. Their size is of 80 to 120 nm. In diameter. The molecular mass averages 40,000 kDa. A nucleocapsids protect the virus when it is outside the host cell.

The spikes are the most distinguishing feature of corona viruses and are responsible for the corona- or halo-like surface. On average a coronavirus particle has 74 spikes.[53] Each spike is about 20 nm long .The virus binds and fuses with the host cell under the action of proteases such as cathepsin family.

Genome: SARS-CoV genome

Coronaviruses contain a single-stranded RNA genome. The genome size ranges from 26.4 to 31.7 kilobases, one of the largest among RNA viruses.

CELL ENTRY AND RELEASE

Infection begins when the viral spike protein attaches to its complementary host cell receptor. Then, a protease of the host cell cleaves the spike protein. Cleavage allows the virus to enter the host cell by endocytosis or direct fusion of the viral envelope with the host membrane.

The replicated genomic RNA becomes the genome of the progeny viruses. They are then released from the host cell by exocytosis through secretory vesicles. Once released the viruses can infect other host cells.

ORIGINS

The ancestors of the alpha, beta, gamma and delta corona viruses were supposed be between 3300 and 2400 BCE.

Bats and birds, as warm-blooded flying vertebrates, are an ideal natural reservoir for the coronavirus gene pool. The global range of bat and avian species that host viruses have enabled extensive dissemination of corona viruses. Many human coronaviruses have their origin in bats. The most closely related bat coronavirus and SARS-CoV diverged in 1986. The ancestors of SARS-CoV first infected leaf-nose bats of the genus *Hipposideridae*; subsequently, they spread to horse-shoe bats in the species *Rhinolophidae*, then to Asian palm civets, and finally to humans. Unlike other betacoronaviruses, bovine coronavirus is thought to have originated in rodents and not in bats. Besides causing respiratory infections, human coronavirus OC43 may also play a role in neurological diseases.Human coronavirus HKU1, like some others also has its origins in rodents.

Infection in humans. SARS-CoV-2 causing COVID-19

Coronaviruses vary significantly in risk factor. Some can kill more than 30% of those infected, such as MERS-CoV, and some are relatively harmless; some cause colds with major symptoms, such as fever, and a sore throat and pneumonia. The SARS-CoV[severe acute respiratory syndrome], uniquely t causes both upper and lower respiratory tract infections.

Six species of human coronaviruses are known, with one species subdivided into two different strains, making seven, prefererentially detectable during November to March.

Four are generally mild, which might have been aggressive in the past:

1. Human coronavirus OC43 (HCoV-OC43), β-CoV
2. Human coronavirus HKU1 (HCoV-HKU1), β-CoV
3. Human coronavirus 229E (HCoV-229E), α-CoV
4. Human coronavirus NL63 (HCoV-NL63), α-CoV–

Three human coronaviruses produce potentially severe symptoms:

1. Severe acute respiratory syndrome coronavirus (SARS-CoV), β-CoV (identified in 2003)
2. Middle East respiratory syndrome-related coronavirus (MERS-CoV), β-CoV (identified in 2012)
3. Severe acute respiratory syndrome coronavirus 2 (SARS-CoV-2), β-CoV (identified in 2019)

These cause the diseases commonly called SARS, MERS, and COVID-19 respectively.

COMMON COLD

Although the common cold is usually caused by rhinoviruses,

in about 15% of cases the cause is a coronavirus. The human coronaviruses HCoV-OC43, HCoV-HKU1, HCoV-229E, and HCoV-NL63 continually circulate in the human population in adults and children worldwide and produce the generally mild symptoms of the common cold. The four mild coronaviruses have a seasonal incidence occurring in the winter months in temperate climates. There is no preponderance in any season in tropical climates.

In 2003, following the outbreak of severe acute respiratory syndrome (SARS) which had begun the prior year in Asia, and secondary cases elsewhere in the world, WHO issued a press release stating that a novel coronavirus identified by several laboratories was the causative agent for SARS. The virus was officially named the SARS coronavirus (SARS-CoV). More than 8,000 people from 29 different countries and territories were infected, and at least 774 died.

MIDDLE EAST RESPIRATORY SYNDROME (MERS)

In September 2012, a new type of coronavirus was identified, officially named Middle East respiratory syndrome coronavirus (MERS-CoV. The WHO update on 28 September 2012 said the virus did not seem to pass easily from person to person.

However, human-to-human transmissions were reported in in France, Tunisia, Qatar and Saudi Arabia. By 30 October 2013, there were 124 cases and 52 deaths in Saudi Arabia.The final name for the virus is Middle East respiratory syndrome coronavirus (MERS- CoV).

In May 2015, an outbreak of MERS-CoV occurred in the Republic of Korea, when a man who had traveled to the Middle East, visited four hospitals in the Seoul area to treat his illness. This caused one of the largest outbreaks of MERS-CoV outside the Middle East.[116] As of December 2019, 2,468 cases of MERS-CoV infection had been confirmed by laboratory tests, 851 of which were fatal, a mortality rate of approximately 34.5%

CORONAVIRUS DISEASE 2019 (COVID-19)

In December 2019, a pneumonia outbreak was reported in Wuhan, China. On 31 December 2019, the outbreak was traced to a novel strain of coronavirus, which was given the interim name 2019-nCoV by the W H O later renamed SARS-CoV-2.

As of 9 March 2022, there have been at least 6,017,018 confirmed deaths and more than 450,174,901 confirmed cases in the COVID-19 pandemic. The Wuhan strain has been identified as a new strain of Betacoronavirus from group 2B with approximately 70% genetic similarity to the SARS-CoV. The virus has a 96% similarity to a bat coronavirus, so it is widely suspected to originate from bats as well.

INFECTION IN ANIMALS

Coronaviruses have known in veterinary medicine since the 1930s, infecting swine, cattle, horses, camels, cats, dogs, rodents, birds and bats. The majority of animal related coronaviruses infect the intestinal tract and are transmitted by a fecal-oral route.

Coronaviruses infect domesticated birds, such as Chicken, pig farming and the Cattle raising, Turkeys, **Domestic pets** such as cats, dogs, and ferrets and **Laboratory animals such as mice, rats, rabbits.**

Characteristics of zoonotic coronavirus strains MERS-CoV, SARS-CoV, SARS-CoV-2,and related diseases

Characteristics of zoonotic coronavirus strains MERS-CoV, SARS-CoV, SARS-CoV-2,and related diseases			
	MERS-CoV	SARS-CoV	SARS-CoV-2
Disease	MERS	SARS	COVID-19
Outbreaks	2012, 2015, 2018	2002–2004	2019–2021 pandemic
Epidemiology			
Date of first identified case	June 2012	November 2002	December 2019[98]
Location of first identified case	Jeddah, Saudi Arabia	Shunde, China	Wuhan, China
Age average	56	44[99][a]	56[100]
Sex ratio (M:F)	3.3:1	0.8:1[101]	1.6:1[100]
Confirmed cases	2494	8096[102]	450,174,901[103][b]
Deaths	858	774[102]	6,017,018[103][b]
Case fatality rate	37%	9.2%	1.34%[103]
Symptoms			
Fever	98%	99–100%	87.9%[104]
Dry cough	47%	29–75%	67.7%[104]
Dyspnea	72%	40–42%	18.6%[104]
Diarrhea	26%	20–25%	3.7%[104]
Sore throat	21%	13–25%	13.9%[104]
Ventilatory use	24.5%[105]	14–20%	4.1%[106]

COVID-19A

SARS-CoV-2

INTRODUCTION

Coronavirus disease 2019 (COVID-19) is a contagious disease caused by severe acute respiratory syndrome coronavirus 2 (SARS-CoV-2). The first known case was identified in Wuhan, China, in December 2019.[7] The disease has since spread worldwide, leading to an ongoing pandemic.

Symptoms of COVID-19 are variable, covering headache,), nasal congestion, runny nose, cough, muscle pain, sore throat, fever, diarrhea, breathing difficulties, fatigue, breathing difficulties, loss of smell (anosmia) and taste (ageusia). People with the same infection may have different symptoms, and their symptoms may change over time.

Symptoms may begin 1 to 14 days after exposure to the virus. At least a third of people who are infected do not develop them. Of those people who develop symptoms, most (81%) develop moderate symptoms (up to mild pneumonia), while 14% develop severe symptoms (dyspnea, hypoxia, or more than 50% lung involvement on imaging), and 5% suffer critical symptoms (respiratory failure, shock, or multiorgan dysfunction).

Older people are at a higher risk of developing severe symptoms. Some people continue to experience a range of effects (long COVID) for months after recovery, and damage to organs has been observed.

TRANSMISSION

COVID-19 is mainly transmitted when people breathe in air contaminated by the virus. Infected people exhale those particles as they breathe, talk, cough, sneeze, or sing. The risk of breathing these in is highest when people are in close proximity, but they can be inhaled over longer distances, particularly indoors. Transmission can also occur if splashed or sprayed with contaminated fluids in the eyes, nose or mouth, and, rarely, via contaminated surfaces. People remain contagious for up to 20 days, and can spread the virus even if they do not develop symptoms.

NOMENCLATURE

During the initial outbreak in Wuhan, the virus and disease were referred to as "coronavirus" and "Wuhan coronavirus", with the disease called "Wuhan pneumonia". In the past, many diseases have been named after geographical locations, such as the Spanish flu, But now WHO recommended the official names COVID19 and SARS-CoV-2.

VIROLOGY

SARS-CoV-2 is a severe acute respiratory syndrome coronavirus. It was first isolated from three people with pneumonia in Wuhan. All structural features of the novel SARS-CoV-2 virus particle occur in related coronaviruses in nature. SARS-CoV-2 is closely related to the original SARS-CoV. It is thought to have an animal (zoonotic) origin.

SARS-COV-2 VARIANTS

The many thousands of SARS-CoV-2 variants are grouped into either clades or lineages. The WHO, in collaboration with partners, expert networks, national authorities, institutions and researchers, have established nomenclature systems for naming and tracking SARS-CoV-2 At the present time, the expert group convened by WHO has recommended the labeling of variants using letters of the Greek Alphabet, for example, Alpha, Beta, Delta, and Gamma, giving the justification that they "will be easier and more practical to be discussed by non-scientific audiences."As of December 2021, there are five dominant variants of SARS-CoV-2 spreading among global populations: the Alpha variant (UK variant), the Beta variant (South Africa variant), the Gamma variant{ Brazil variant), the Delta variant (India variant), and the Omicron variant , which had spread to 57 countries as of 7 December

IMPACTS ON HEALTH

RESPIRATORY SYSTEM

Following viral entry, COVID19 infects the ciliated epithelium of the naso pharynx and upper airways.

NERVOUS SYSTEM

One common symptom, loss of smell, results from infection of the support cells of the olfactory epithelium, with subsequent damage to the olfactory neurons. The involvement of both the central and peripheral nervous system in COVID19 has been reported. It is clear that many people with COVID-19 exhibit neurological or mental health issues. COVID-19 can cause brain damage, whether temporary of permanent is unclear.

GASTROINTESTINAL SYSTEM

The virus also affects gastrointestinal system as ACE2 is abundantly shown in the glandular cells of gastric, duodenal and rectal epithelium as well as endothelial cells and enterocytes of the small intestine.

CARDIOVASCULAR SYSTEM

The virus can cause acute myocardial injury and chronic damage to the cardiovascular system. Such an injury was found in 12% of infected people in an hospital in Wuhan, China.

A high incidence of thrombosis and venous thromboembolism have been found in people transferred to Intensive care units after infections. Blood vessel dysfunction and clot formation play a significant role in mortality. Incidences of clots leading to pulmonary embolisms, and ischaemic events within the brain have been noted as complications leading to death.

OTHER ORGANS

Another common cause of death is complications related to the kidneys. Early reports show that up to 30% of hospitalized patients both in China and in New York have experienced some injury to their kidneys, including some persons with no previous kidney problems.

LONG TERM EFFECTS

A majority of those who were admitted to hospital with severe disease report long-term problems, including fatigue and shortness of breath. By a variety of mechanisms, the lungs are the organs most affected in COVID 19. People with advanced age, severe disease, prolonged

ICU stays, or who smoke are more likely to have long-lasting effects, including pulmonary fibrosis.

IMMUNITY

The immune response by humans to SARS-CoV-2 virus occurs as a combination of the cell-mediated immunity and antibody production. Reinfection with COVID19 is possible but uncommon. A systematic review found 17 cases of confirmed reinfection in medical literature.

MORTALITY

The mortality rate reflects the number of deaths within a specific demographic group divided by the population of that demographic group. Consequently, the mortality rate reflects the prevalence as well as the severity of the disease within a given population. Mortality rates are highly correlated to age, with relatively low rates for young people and relatively high rates among the elderly. fact, one relevant factor of mortality rates is the age structure of the countries' population. For example, the fatality rate for COVID19 is lower in India than in the US since India's younger population represents a larger percentage than in the US.

CASE FATALITY RATE

The case fatality rate (CFR) reflects the number of deaths divided by the number of diagnosed cases within a given time interval. Based on Johns Hopkins University statistics, the global death-to-case ratio is 1.34% (6,017,018/450,174,901) as of 9 March 2022. The number varies by region.

INFECTION FATALITY RATE

A key metric in gauging the severity of COVID19 is the infection fatality rate (IFR), also referred to as the *infection fatality ratio* or *infection fatality risk*. This metric is calculated by dividing the total number of deaths from the disease by the total number of infected individuals; hence, in contrast to the CFR, the IFR incorporates asymptomatic and undiagnosed infections as well as reported cases

SEX DIFFERENCES

COVID 19 case fatality rates are higher among men in most countries except India, Nepal, Vietnam, and Slovenia. One meta-analysis found that globally, men were more likely to get COVID19 than women; there were approximately 55 men and 45 women per 100 infections.

The Chinese reported the death rate was 2.8% for men and 1.7% for women. Later reviews in June 2020 indicated that there is no significant difference in susceptibility or in CFR between genders. It may be attributable to lifestyle choices such as smoking and drinking alcohol rather than genetic factors. Smoking, which in some countries is mainly a male activity, is a habit that contributes to fatality rates among men Sex-based immunological differences, In Europe as of February 2020, 57% of the infected people were men and 72% of those died with COVID19 were men.

Viral illnesses like Ebola, HIV, influenza and SARS affect men and women differently.

ETHNIC DIFFERENCES

In the US, a greater proportion of deaths due to COVID19 have occurred among African Americans and other minority groups.

Structural factors that prevent them from practicing social distancing include their concentration in crowded substandard housing and in "essential" occupations such as retail grocery workers, public transit employees, health-care workers and custodial staff. Greater prevalence of lacking health insurance and care of underlying conditions such as diabetes, hypertension, and heart disease also increase their risk of death.

Similar issues affect Native American and Latino communities.

According to a US health policy non-profit, 34% of American Indian and Alaska Native People (AIAN) non-elderly adults are at risk of serious illness compared to 21% of white non-elderly adults. The source attributes it to disproportionately high rates of many health conditions that may put them at higher risk as well as living conditions like lack of access to clean water.

A greater proportion of deaths due to COVID19 have occurred in those of a Black, Asian, and other ethnic minority background. More severe impacts upon victims including the relative incidence of the necessity of hospitalization requirements, and vulnerability to the disease has been associated via DNA analysis to be expressed in genetic variants at chromosomal region 3, features that are associated with European Neanderthal heritage.

CO-MORBIDITIES

Most of those who die of COVID19 have pre-existing conditions, such as hypertension, diabetes mellitus, and cardiovascular disease.

According to this report the most common comorbidities are hypertension (66% of deaths), type 2 diabetes (29.8% of deaths), Ischemic Heart Disease (27.6% of deaths), atrial fibrillation (23.1% of deaths) and chronic renal failure (20.2% of deaths). Most critical respiratory comorbidities according to the Centers for Disease Control and Prevention (CDC), are: moderate or severe asthma, pulmonary fibrosis, cystic fibrosis.

HISTORY

The virus is thought to be of animal origin.. There are several theories about the index case origin and investigations into the origin of the pandemic are ongoing. Phylogenetics estimates that SARS-CoV-2 arose in October or November 2019 and may have been circulating in Guangdong before. Evidence suggests that it descends from a coronavirus that infects wild bats, and spread to humans. The possibility that the virus was accidentally released from a laboratory is also under consideration. USA found that the virus was not developed as a biological weapon and that it is unlikely to have been genetically engineered.

The first confirmed human infections were in Wuhan, Hubei, China. In January 2020 *The Lancet* report on first 41 cases fixed the earliest date of onset as 1 December 2019. WHO reported the earliest onset of symptoms as 8 December 2019. Traces of the virus have been found in wastewater samples that were collected in Milan and Turin, Italy, on 18 December 2019. Human-to-human transmission was confirmed by the WHO and Chinese authorities by 20 January 2020. These were mostly linked to the Hunan Seafood Wholesale Market. By December 2019, the spread of infection was almost entirely driven by human-to-human transmission.

Hubei and Wuhan increasingly showed infections. Doctors and lab analysts suspected pneumonia – like infection – making them release "the treatment of pneumonia of unknown cause". But eight of these doctors, were later admonished by the police for spreading false rumours and raising the alarm !

The Wuhan Municipal Health Commission made the first public announcement of a pneumonia outbreak of unknown cause on 31 December, confirming 27 cases - enough to trigger an investigation.

During the early stages, the number of cases doubled weekly . In early and mid-January 2020, the virus spread to other provinces, helped by the New Year migration and Wuhan being a transport hub and major rail interchange.

A report in *The Lancet* on 24 January indicated human transmission, strongly recommended personal protective equipment for health workers, and said testing for the virus was essential due to its "pandemic potential". On 30 January, the WHO declared COVID-19 a Public Health Emergency of International Concern.

Initial deaths occurred in Italy and overtook China, and both by the USA with in a few weeks. Gradually most world was engulfed by this endemic turned Pandemic . As of 1 October 2021, *Reuters* reported that it had estimated the worldwide total number of deaths due to COVID19 to have exceeded five million.

IMPACT OF THE COVID-19 PANDEMIC ON ANIMALS

Humans appear to be capable of spreading the virus to some other animals, a type of disease transmission referred to as zooanthroponosis. Cats and ferrets, can catch this virus from infected humans. Symptoms in cats include respiratory and digestive . Cats can spread the virus to other cats, and to humans. Dogs are less susceptible to this infection. Behaviors which increase the risk of transmission include kissing, licking, and petting the animal.

The virus does not appear to be able to infect pigs, ducks, or chickens at all. Mice, rats, and rabbits, if they can be infected at all, are unlikely to be involved in spreading the virus. Tigers and lions in zoos have become infected as a result of contact with infected humans. As expected, monkeys and great ape species such as orangutans can also be infected with the COVID19 virus.

Minks have been infected. Minks may be asymptomatic, and can also spread the virus to humans. Multiple countries have identified infected animals in mink farms. Denmark, a major producer of mink pelts, ordered the slaughter of all minks over fears of viral mutations, following an outbreak.

BIOETHICS

Since the outbreak of the COVID19 pandemic, scholars have explored the bioethics, normative economics, and political theories of healthcare policies related to the public health crisis. Academics have pointed to the moral distress of healthcare workers, ethics of distributing scarce healthcare resources such as ventilators, and the global justice of vaccine diplomacies. The socio-economic inequalities between genders, races, groups with disabilities, communities, regions, countries, and continents have also drawn attention in academia and the general public.

MUTATIONS

OVERVIEW

In biology, a **mutation** is an alteration in the nucleotide sequence of the genome of an organism. Mutations result from errors during DNA mitosis, or meiosis or other types of damage to DNA, which then may undergo error-prone repair, cause an error during other forms of repair, or cause an error during replication. Mutations may also result from insertion or deletion of segments of DNA due to mobile genetic elements.

Mutations may or may not produce detectable changes in the observable characteristicsof an organism. Mutations play a part in both normal and abnormal biological processes including: evolution, cancer and the development of the immune system, including functional diversity. Mutation is the ultimate source of all genetic variation, providing the raw material on which evolutionary forces such as natural selection can act.

Mutation can result in many different types of change in sequences. Mutations in genes can have no effect, alter the product of a gene, or prevent the gene from functioning properly or completely. Mutations can also occur in nongenic regions. A 2007 study on genetic variations between different species of *Drosophila* suggested that, if a mutation changes a protein produced by a gene, the result is likely to be harmful, with an estimated 70% of amino acid polymorphisms that have damaging effects, and the remainder being either neutral or marginally beneficial. Due to the damaging effects that mutations can have on genes, organisms have mechanisms such as DNA repair to prevent or correct mutations by reverting the mutated sequence back to its original state.

Mutations can involve the duplication of large sections of DNA, usually through genetic recombination. These duplications are a major source of raw material for evolving new genes, with tens to hundreds of genes duplicated in animal genomes every million years. Most genes belong to larger gene families of shared ancestry, detectable by their sequence homology. Novel genes are produced by several methods, commonly through the duplication and mutation of an ancestral gene, or by recombining parts of different genes to form new combinations with new functions.

Changes in chromosome number may involve even larger mutations, where segments of the DNA within chromosomes break and then rearrange. For example, in the Hominidae, two chromosomes fused to produce human chromosome 2; this fusion did not occur in the lineage of the other apes, and they retain these separate chromosomes. In evolution, the most important role of such chromosomal rearrangements may be to accelerate the divergence of a population into new species by making populations less likely to interbreed, thereby preserving genetic differences between these populations.

Nonlethal mutations accumulate within the gene pool and increase the amount of genetic variation. The abundance of some genetic changes within the gene pool can be reduced by natural selection, while other "more favorable" mutations may accumulate and result in adaptive changes.

Neutral mutations are defined as mutations whose effects do not influence the fitness of an individual. These can increase in frequency over time due to genetic drift. It is believed that the overwhelming majority of mutations have no significant effect on an organism's fitness. Also, DNA repair mechanisms are able to mend most changes before they become permanent mutations, and many organisms have mechanisms for eliminating otherwise-permanently mutated somatic cells.

Humans on average pass 60 new mutations to their children but fathers pass more mutations depending on their age with every year adding two new mutations to a child.

Beneficial mutations can improve reproductive success.

CAUSES

Four classes of mutations are (1) spontaneous mutations, (2) mutations due to error-prone replication naturally occurring DNA damage (3) errors introduced during DNA repair, and (4) induced mutations caused by mutagens.

Scientists may also deliberately introduce mutant sequences through DNA manipulation for the sake of scientific experimentation.

One 2017 study claimed that 66% of cancer-causing mutations are random, 29% are due to the environment (the studied population spanned 69 countries), and 5% are inherited.

SPONTANEOUS MUTATION

Spontaneous mutations occur with non-zero probability even given a healthy, uncontaminated cell. Naturally occurring oxidative DNA damage is estimated to occur 10,000 times per cell per day in humans and 100,000 times per cell per day in rats. Spontaneous mutations can be characterized by the specific change:

ERROR-PRONE REPLICATION

There is increasing evidence that the majority of spontaneously arising mutations are due to past DNA damage in the template strand.

ERRORS INTRODUCED DURING DNA REPAIR

Although naturally occurring double-strand breaks occur at a relatively low frequency in DNA, their repair often causes mutation.

INDUCED MUTATION

Induced mutations are alterations in the gene after it has come in contact with mutagens and environmental causes.

CHEMICALS

Hydroxylamine, Base analogs (e.g., Bromodeoxyuridine (BrdU))

Alkylating agents (e.g., *N*-ethyl-*N*-nitrosourea (ENU).

These agents can mutate both replicating and non-replicating DNA. In contrast, a base analog can mutate the DNA only when the analog is incorporated in a replicating DNA. Each of these classes of chemical mutagens has certain effects that then lead to transitions, transversions, or deletions.

RADIATION

Ultraviolet light (UV) (including non-ionizing radiation).

Two nucleotide bases in DNA—cytosine and thymine—are most vulnerable to radiation that can change their properties. UV light can induce adjacent pyrimidine bases in a DNA strand to become covalently joined as a pyrimidine dimer. UV radiation, in particular longer-wave UVA, can also cause oxidative damage to DNA.

IONIZING RADIATION

Exposure to ionizing radiation, such as gamma radiation, can result in mutation, possibly resulting in cancer or death.

CLASSIFICATION OF TYPES

By effect on structure

Five types of chromosomal mutations

The sequence of a gene can be altered in a number of ways. Gene mutations have varying effects on health depending on where they occur and whether they alter the function of essential proteins. Mutations in the structure of genes can be classified into several types.

Large-scale mutations in chromosomal structure include:

Amplifications or gene duplications. Large scale changes to the structure of chromosomes called chromosomal rearrangement that can lead to a decrease of fitness but also to speciation in isolated, inbred populations. These include: Chromosomal translocations: interchange of genetic parts from nonhomologous chromosomes. Chromosomal inversions: reversing the orientation of a chromosomal segment.

Non-homologous chromosomal crossover. Loss of heterozygosity: loss of one allele, either by a deletion or a genetic recombination event, in an organism that previously had two different alleles.

SMALL-SCALE MUTATIONS

Small-scale mutations affect a gene in one or a few nucleotides. If only a single nucleotide is affected, they are called point mutations.

Deletions remove one or more nucleotides from the DNA. Like insertions, these mutations can alter the reading frame of the gene. In general, they are irreversible: Though exactly the same sequence might, in theory, be restored by another insertion,

Substitution mutations, often caused by chemicals or malfunction of DNA replication, exchange a single nucleotide for another.

These changes are classified as transitions or transversions. Most common is the transition that exchanges a purine for a purine or a pyrimidine for a pyrimidine, A transition can be caused by nitrous acid, base mispairing, or mutagenic base analogs such as BrdU. Less common is a transversion, which exchanges a purine for a pyrimidine or a pyrimidine for a purine. An example of a transversion is the conversion of adenine into a cytosine. Point mutations are modifications of single base pairs of DNA or other small base pairs within a gene.

A frameshift mutation is caused by insertion or deletion of a number of nucleotides that is not evenly divisible by three from a DNA sequence. Due to the triplet nature of gene expression by codons, the insertion or deletion can disrupt the reading frame, or the grouping of the codons, resulting in a completely different translation from the original. The earlier in the sequence the deletion or insertion occurs, the more altered the protein produced is.

BY EFFECT ON FUNCTION

By effect on fitness : It is sometimes useful to classify mutations as **harmful or beneficial** or **neutral**

A harmful, or deleterious, mutation decreases the fitness of the organism. Many, but not all mutations in essential genes are harmful (if a mutation does not change the amino acid sequence in an essential protein, it is harmless in most cases).

A beneficial, or advantageous mutation increases the fitness of the organism. Examples are mutations that lead to antibiotic resistance in bacteria (which are beneficial for bacteria but usually not for humans).

A neutral mutation has no harmful or beneficial effect on the organism. Such mutations occur at a steady rate, forming the basis for the molecular clock. In the neutral theory of molecular evolution, neutral mutations provide genetic drift as the basis for most variation at the molecular level. In animals or plants, most mutations are neutral, given that the vast majority of their genomes is either non-coding or consists of repetitive sequences that have no obvious function ("junk DNA").

BY INHERITANCE

A mutation has caused this moss rose plant to produce flowers of different colors. This is a somatic mutation that may also be passed on in the germline.

In multicellular organisms with dedicated reproductive cells, mutations can be subdivided into germline mutations, which can be passed on to descendants through their reproductive cells, and somatic mutations (also called acquired mutations), which involve cells outside the dedicated reproductive group and which are not usually transmitted to descendants.

Diploid organisms (e.g., humans) contain two copies of each gene—a paternal and a maternal allele. Based on the occurrence of mutation on each chromosome, we may classify mutations into three types. A wild type or homozygous non-mutated organism is one in which neither allele is mutated.

A heterozygous mutation is a mutation of only one allele.

GERMLINE MUTATION

A germline mutation in the reproductive cells of an individual gives rise to a *constitutional mutation* in the offspring, that is, a mutation that is present in every cell. A constitutional mutation can also occur very soon after fertilisation, or continue from a previous constitutional mutation in a parent. A germline mutation can be passed down through subsequent generations of organisms.

The distinction between germline and somatic mutations is important in animals that have a dedicated germline to produce reproductive cells. However, it is of little value in plants. The distinction is also blurred in those animals that reproduce asexually through mechanisms such as budding, because the cells that give rise to the daughter organisms also give rise to that organism's germline. A new germline mutation not inherited from either parent is called a *de novo* **mutation.**

SOMATIC MUTATION

A change in the genetic structure that is not inherited from a parent, and also not passed to offspring, is called a somatic mutation. Somatic mutations are not inherited by an organism's offspring because they do not affect the germline. However, they are passed down to all the progeny of a mutated cell within the same organism during mitosis. A major section of an organism therefore might carry the same mutation. These types of mutations are usually prompted by environmental causes, such as ultraviolet radiation or any exposure to certain harmful chemicals, and can cause diseases including cancer.

With plants, some somatic mutations can be propagated without the need for seed production, for example, by grafting and stem cuttings. These type of mutation have led to new types of fruits, such as the "Delicious" apple.

NOMENCLATURE

The committee of the Human Genome Variation Society (HGVS) has developed the standard human sequence variant nomenclature, which should be used by researchers and DNA diagnostic centers to generate unambiguous mutation descriptions. In principle, this nomenclature can also be used to describe mutations in other organisms. The nomenclature specifies the type of mutation and base or amino acid changes. If it becomes necessary to differentiate between mutations in genomic DNA, mito-chondrial DNA, and RNA, a simple convention is used, which however is not relevant here.

RANDOMNESS, RECESSIVENES, REPEATEDNESS AND REVERSIBILITY

There is a widespread assumption that mutations are "random"

And cannot be directed. But this cannot be entirely true. For instance, biologically important regions were found to be protected from mutations and mutations beneficial to the studied plant were found to be more likely – i.e. mutation is "non-random in a way that benefits the plant"

Most mutations are recessive and therefore remain in a heterozygous state. They are expressed only when they become homozygous. Same locus mutates repeatedly from one to another form and character. Because of this character, often a mutant reverse to its original in a few generations. This happens generally in microbes. Thus a streptomycin resistant strain becomes sensitive and dies off. Some generations latter the sensitive one became dependent and resistant in due course!

DISEASE CAUSATION

Changes in DNA caused by mutation in a coding region of DNA can cause errors in protein sequence that may result in partially or completely non-functional proteins. Each cell, in order to function correctly, depends on thousands of proteins to function correctly. When a mutation alters a protein that plays a critical role in the body, a medical condition resuls.

One study on the comparison of genes in *Drosophila* suggests that if a mutation does change a protein, the mutation will harmful. An estimated 70 percent of amino acid polymorphisms have damaging effects, and the remainder being either neutral or weakly beneficial. Some mutations alter a gene's DNA base sequence but do not change the protein made by the gene. Studies have shown that only 7% of point mutations in noncoding DNA and 12% in coding DNA are deleterious in yeast. The rest of the mutations are either neutral or slightly beneficial.

INHERITED DISORDERS

If a mutation is present in a germ cell, it can give rise to offspring that carries the mutation in all of its cells. This is the case in hereditary diseases. In particular, if there is a mutation in a DNA repair gene within a germ cell, humans carrying such germline mutations may have an increased risk of cancer. 34 such germline mutations are so far known. An example is albinism, Individuals with this disorder are more prone to many types of cancers, other disorders and have impaired vision.

DNA damages are repaired by the DNA repair system of the cell. Each cell has a number of pathways through which enzymes recognize and repair damages in DNA. Because DNA can be damaged in many ways, the process of DNA repair is an important way in which the body protects itself from disease. Once DNA damage has given rise to a mutation, the mutation cannot be repaired.

ROLE IN CARCINOGENESIS

On the other hand, a mutation may occur in a somatic cell of an organism. Such mutations will be present in all descendants of this cell within the same organism. The accumulation of certain mutations over generations of somatic cells is part of cause of malignant transformation, from normal cell to cancer cell.

Cells with heterozygous loss-of-function mutations (one good copy of gene and one mutated copy) may function normally until the good copy has been spontaneously mutated. This kind of mutation happens often in living organisms.

Point mutations may arise from spontaneous mutations that occur during DNA replication. The rate of mutation may be increased by mutagens. Mutagens can be physical, such as radiation from UV rays, X-rays or extreme heat, or chemical molecules that misplace base pairs . Mutagens associated with cancers are often studied to learn about cancer and its prevention.

BENEFICIAL MUTATIONS

Although mutations that cause changes in protein sequences can be harmful to an organism, on occasions the effect may be positive in a given environ-ment. In this case, the mutation may enable the mutant organism to withstand particular environmental stresses better than wild-type organisms, or reproduce more quickly. In these cases a mutation will tend to become more common in a population through natural selection. Examples include the following:

HIV RESISTANCE

A specific 32 base pair deletion in human CCR5 confers HIV resistance to homozygotes and delays AIDS onset in heterozygotes.

One possible explanation of the etiology of the relatively high frequency of CCR5-Δ32 in the European population is that it conferred resistance to the bubonic plague in mid-14th century Europe. People with this mutation were more likely to survive infection; thus its frequency in the population increased.[110] This theory could explain why this mutation is not found in Southern Africa, which remained untouched by bubonic plague.

A newer theory suggests that the selective pressure on the CCR5 Delta 32 mutation was caused by smallpox instead of the bubonic plague.

MALARIA RESISTANCE

An example of a harmful mutation is sickle-cell disease, a blood disorder in which the body produces an abnormal type of the oxygen-carrying substance hemoglobin in the red blood cells. One-third of all indigenous inhabitants of Sub-Saharan Africa carry the allele, because, in areas where malaria is common, there is a survival value in carrying only a single sickle-cell allele (sickle cell trait). Those with only one of the two alleles of the sickle-cell disease are more resistant to malaria, since the infestation of the malaria *Plasmodium* is halted by the sickling of the cells that it infests.

ANTIBIOTIC RESISTANCE

Practically all bacteria develop antibiotic resistance when exposed to antibiotics. In fact, bacterial populations already have such mutations that get selected under antibiotic selection. Obviously, such mutations are only beneficial for the bacteria but not for those infected.

LACTASE PERSISTENCE

A mutation allowed humans to express the enzyme lactase after they are naturally weaned from breast milk, allowing adults to digest lactose, which is likely one of the most beneficial mutations in recent human evolution.

HISTORY

Mutationism is one of several alternatives to Darwinian evolution that have existed both before and after the publication of Darwin's 1859 book, *On the Origin of Species*. In the theory, mutation was the source of novelty, creating new forms and new species, potentially instantaneously, in a sudden jump. This was envisaged as driving evolution, which was limited by the supply of mutations.

Before Darwin, biologists commonly believed in saltationism, the possibility of large evolutionary jumps, including immediate speciation. For example, in 1822 Saint-Hilaire argued that species could be formed by sudden transformations, or what would later be called macromutation. Darwin opposed saltation, insisting on gradualism in evolution as in geology. In 1864, Kölliker revived

Geoffroy's theory. In 1901 the geneticist Hugo de Vries gave the name "mutation" to seemingly new forms that suddenly arose in his experiments on the evening primrose *Oenothera lamarckiana*, and in the first decade of the 20th century, mutationism, or as de Vries named it m*utationstheorie*, became a rival to Darwinism supported for a while by geneticists including Bateson, t Morgan, and Punnett.

Understanding of mutationism is clouded by the mid-20th century portrayal of the early mutationists by supporters of the modern synthesis as opponents of Darwinian evolution and rivals of the biometrics school who argued that selection operated on continuous variation. In this portrayal, mutationism was defeated by a synthesis of genetics and natural selection that supposedly started later, around 1918, with work by the mathematician Ronald Fisher. However, the alignment of Mendelian genetics and natural selection began as early as 1902 with a paper by Yule, and built up with theoretical and experimental work in Europe and America. Despite the controversy, the early mutationists had by 1918 already accepted natural selection and explained continuous variation as the result of multiple genes acting on the same characteristic, such as height.

Mutationism, along with other alternatives to Darwinism like Lamarckism and orthogenesis, was discarded by most biologists as they came to see that Mendelian genetics and natural selection could readily work together; mutation took its place as a source of the genetic variation essential for natural selection to work on.

However, mutationism did not entirely vanish. In 1940, Richard Goldschmidt again argued for single-step speciation by macro-mutation, describing the organisms thus produced as "hopeful monsters", earning widespread ridicule. In 1987, Masatoshi Nei argued controversially that evolution was often mutation-limited.

ENVIRONMENTAL MUTAGENESIS

From a variety of issues and aspects covered on mutations, we are now concerned only on the spontaneous mutations. Here a specialist is not directly involved in the induction of any mutation. In fact it is not known whether mutations are induced. But all are alerted after experiencing the impacts of mutations. We go backward to origins after the impacts. The science here is deductive.

So much is known about this aspect of mutations in the post 2nd WW world that scattered efforts of scientists had to be organized. The entire effort initiated from ORNL was christened as Environmental Mutagenesis and a Society was organized from 1969. Human Society is directly involved both in causing them as well as suffering.

We are concerned with human ill health appearing irreversible, even the medical practices appearing helpless, but the causative agents spread everywhere and do not appear to wane. Heritable diseases, cancers, birth defects, curability difficulties of some diseases, newer epidemics - these amounting to be about 15-20% of the world. Chemicals known and unknown are uncountable. Pollutions by them are unchartable. EMSUSA and the same from other countries have been making efforts to alleviate the situation.

Complexities are multiplied with Covid, because similar new endemics or pandemics can crop up over above the 1996 Scenario presented by a Journalist in her book "The Coming Plague" The entire field, half a century young, is presented by me pictorially to avoid excessive verbiage in this chapter. Prior to this, I attempt to give a summary of scientific efforts during the 5 decades.

Many national and international societies specialize in genetic toxicology; a link to most of them can be found. One should consider joining those relevant to your location and attending their meetings, partly to obtain the latest information on test methods and strategy, and partly to network with experts in the individual test methods who may be able to help you with assay development, training, or trouble shooting. I found the following groups of particular interest:

GENEPIDEMICS

Human health is impaired beyond human endurance limits. And this happens due to mutations. This was the import of thoughts of several leading geneticists in the west. They decided to initiate some efforts to mitigate the impacts. Therefore was borne the Environmental Mutagen Society in USA. The initial data confirmed their suspicions. These could be in Epidemic proportions was suspected around 1990 and presented in a global conference in Malaysia, in 1987 and published subsequently. They were again substantiated in 2009 in an ICMR conference at Bangalore [September] and feared that epidemics cold become pandemics.

What constitute Genepidemics ?

DIRECT

Humans

1. Genetic Diseases: have been continuing for long
2. Birth Defects : began to appear and increase after the war
3. Carcinogenesis: Science was increasingly getting evidences that mutations change normal cells into malignant ones.
4. Detriments: Some Scientists [Sankarnarayanan and cziezel] believed that many small changes take place in human bodies due to muta-tions which cumulatively reduce human abilities and competences.

Around 1969, the above were truer but in very low frequency. By 1987 these have not only increased but a new dimension was added by pesticides. They contaminate food chains impacting human societies directly and causing imbalances in agroecosystems through pests, pathogens, microbes and weeds. Thus originated the Genepidemics Theory.

By 2009, following the 150th anniversary of Darwin, a re –look at the above revealed a more poignant situation because of indirect consequences of Mutations. The mutations in pathogens, parasites, vectors and stresses on drug development.

INDIRECT

Parasites, Pathogens, Vectors

Diseases such as Cholera, Malaria, Pneumonia, Diphtheria, Influenza, Pox, Typhoid, Tuberculosis, Shistosomiasis, etc

1. Mutations generating drug resistant strains - Malaria, Tuberculosis
2. Mutations causing more virulent strains - Influenza
3. Harmless viruses or bacteria can mutate and behave like Corona

Among many causes for mutagenesis, Drugs, Foods, Water Pollutions, Pesticides , Industrial Accidents - Bhopal [MIC], Vizag [Styrene] etc, play a greater role In accelerating the mutation frequencies

IMPERATIVES

What to do. I for one was trying to caution scientists and educationist who rule our country about the impending dangers. No one listens. Many more do not want to listen. A few exceptions did exist, Sundaram, Bhatia, Raghu, Chadda, Bhattacharya, - significantly all are from BARC, Bombay.

If corrective steps are initiated with wisdom, as in Europe, America and Japan the imminent dangers can be limited to the level of tolerability. If it continued with complacency, a collective Harakiri cannot be prevented.

Every educational institution, Health systems, regulatory agencies and Industries had to honestly plan and execute a set of DOs and DON'Ts. What are these Dos and DON'Ts? These are already available in Japan, UK and Europe, USA and Canada. For Example accent should be laid on the following

Genetics in Medical Education - Human, Vectors, Parasites, Pathogens,

Microbiology Mutagenesis in Biology and Medical Education

Testing as in USA and EUROPE , Industrial Hygeine

Upgrading the Hospitals and Health systems

Luckily, models for everything needed to do are readily available in the above countries. Unfortunately Scientific personnel are very less and interestingly far less than earlier!

2. THE PRICE OF SILENT MUTATIONS

J. V. Chamary, Laurence D. Hurst

Scientific Amercian, June 1, 2009

Small changes to DNA that were once considered innocuous enough to be ignored are proving to be important in human diseases, evolution and biotechnology

Biologists long thought they understood how genetic mutations cause disease. But recent work has revealed an important twist in the tale and uncovered surprising—even counterintuitive—ways that alterations in DNA can make people sick. The classic view assumed that what are termed "silent" mutations were inconsequential to health, because such changes in DNA would not alter the composition of the proteins encoded by genes. Proteins function in virtually every process carried out by cells, from catalyzing biochemical reactions to recognizing foreign invaders. Hence, the thinking went, if a protein's makeup ends up being correct, any small glitches in the process leading to its construction could not do a body harm.

Yet detective work occasionally traced a disorder to a silent mutation, even though researchers presumed that it could not possibly be the culprit. Similar mysteries popped up in studies of genome evolution, where patterns of changes in.the DNA of various species indicated that many silent mutations were preserved over time—a sign that they were useful to the organisms possessing them. In many species, these changes seemed to help cells make proteins more efficiently, but not in people.

Now investigators are beginning to tease out the effects that silent mutations can have on human health and disease. And findings are suggesting intriguing new avenues for improving the design of genes meant to be used as therapies and for genetic engineering.

SYNONYMOUS BUT NOT THE SAME

How a gene mutation can leave a protein unaffected becomes clear when one looks at the way cells make proteins. The basic formula is simple: a string of DNA nucleotides gives rise to a nearly identical sequence of RNA nucleotides, which in turn is translated into an amino acid chain that folds itself neatly into a protein. The letters of this nucleic acid alphabet are distinguished from one another by their chemical bases—adenine (A), cytosine (C), guanine (G) and thymine (T) in DNA, with uracil (U) substituting for thymine in RNA.

In other words, the instructions encoded in nucleic acids must be converted into the language of amino acids so that their "meaning" (a useful protein) can be expressed. When a gene is thus "expressed," the strands of the DNA double helix separate and cellular machinery transcribes the nucleotide sequence along a single strand into a copy made of RNA. Then this messenger RNA (mRNA) transcript must often be edited into a briefer form before it is ready to be translated into a protein by ribosomes and smaller RNAs called transfer RNA (tRNA). As ribosomes ride along the mRNA, tRNAs arrive to deliver the encoded amino acids. Each tRNA carries a specific amino acid, and most are able to recognize just one particular three nucleotide sequence in the mRNA strand. When a tRNA meets its mRNA match, the ribosome adds the tRNA's amino acid cargo to the growing amino acid chain.

The code that cells use to translate the language of DNA and RNA into protein—the famous genetic code—is thus merely the set of rules that govern which tRNA bears which amino acid. This code has a critical feature: it is redundant. All genes and their mRNAs are organized into three-letter "words." called codons. Sixty-four three-letter codons can be constructed from the four-nucleotide alphabet. Three codons act as stop signals to halt RNA translation, which still leaves 61 possible codons to specify a protein alphabet of only 20 amino acids, so nearly every amino acid is specified by more than one codon. For instance, all codons starting with GG (GGA, GGC,

GGG, GGU) are translated to the amino acid glycine, making those codons synonyms.

Single-letter changes to the DNA, known as point mutations, can therefore change a codon to one that specifies the wrong amino acid (known as a missense mutation) or to a stop signal (nonsense mutation), causing the final protein to be truncated. A single-base change can also alter a stop codon so that it then encodes an amino acid (sense mutation), resulting in a lengthened protein. And a final change is possible: a mutation that alters a nucleotide but yields a synonymous codon. These mutations are the ones termed "silent."

EVIDENCE OF BIAS

Examples certainly abound of the first three types of point mutations having a major impact on human health. Three different point mutations in the genes encoding proteins that make up the hemoglobin molecules in red blood cells are responsible for three separate and grave diseases, for instance. In the case of sickle cell anemia, a missense mutation exchanges a water-loving (hydrophilic) amino acid for a water-avoiding one (hydrophobic), causing the proteins to clump together and produce characteristic sickle-shaped blood cells. In polycythemia disorders, a nonsense mutation truncates one of the hemoglobin proteins, resulting in thickened blood. And in thalassemia, a sense mutation changes a stop codon (TAA) to the codon for glutamine (CAA), creating a much longer and nonfunctional protein.

Only in the 1980s did scientists realize that silent mutations could also affect protein production—at least in bacteria and yeast. A key discovery at the time was that the genes of those organisms did not use synonymous codons in equal numbers. When the bacterium *Escherichia coli* specifies the amino acid asparagine, for instance, the codon AAC appears in its DNA much more often than AAT. The reason for this biased usage of codons soon became apparent: cells were preferentially employing certain codons because those choices enhanced the rate or accuracy of protein synthesis.

It turned out that tRNAs corresponding to those synonymous codons typically are not equally abundant within the cell. Most important, then, a gene that contains more of the codons matching the relatively.abundant tRNAs would be translated faster, because the higher concentration of those tRNAs would make them more likely to be present when needed. In other cases, a single tRNA variety matches more than one synonymous codon but binds more readily to one codon in particular, so the use of that codon maximizes the accuracy of translation. Consequently, a cell has good reasons not to use all codons equally. As expected, in bacteria and yeast the genes that encode especially abundant proteins exhibit the greatest codon bias, with the preferred codons matching the most common or better-binding tRNAs.

Later observations in other organisms—including plants, flies and worms—revealed similar biases. With such a diverse array of species employing this technique to improve the efficiency of protein production, it seemed likely that mammals would, too. Analyses of mammalian genes did indeed reveal tendencies toward favoring certain codons. The similarity between simple organisms and mammals, however, proved to be only superficial. For reasons not yet fully understood, mammalian genomes are organized into large blocks, each with a distinctively skewed nucleotide content: some regions are rich in G and C bases, whereas others are enriched for A and T. As a result, genes residing in a GC-rich region of the genome tend to have many codons contai-ning those bases. Our genes, then, do show a bias for using certain codons, but unlike simpler organisms, the mammalian pattern does not obviously suggest that the reason is to optimize protein synthesis.

For many years these findings seemed to diminish the likelihood that silent mutations influenced the functioning of the human body. Starting in the early 2000s, however, comparisons of the same gene in different species began to hint that this orthodoxy was wrong. One can measure the rate at which gene sequences in two species have diverged by comparing the sites where nucleotides have changed and those where they have remained the same. In principle, any mutation that does not affect an organism is invisible to the force of natural selection, which preserves variations that are beneficial. According to the thinking at the time, regions invisible to selection would include sites of silent mutations within genes as well as the 98 percent of the genome that does not specify protein—the noncoding DNA. Yet when scientists began looking at whether silent sites in genes evolved at the same rate as noncoding regions, they unexpectedly found differences—a sign that silent mutations could affect physiology after all.

BREAKING THE SILENCE

Initially researchers had no idea how such mutations could disturb protein manufacture in mammals. Lately, however, studies of human disease have provided not just one mechanism but many. Silent disease-causing mutations interfere with several stages of the protein-making process, from DNA transcription all the way through to the trans-lation of mRNA into proteins.

One example involves silent mutations changing how a gene transcript is edited. Shortly after a gene is transcribed into RNA form, that transcript is trimmed to remove non-coding regions known as introns. Like a movie editor who cuts out unwanted film, cellular splicing machinery needs to find the good bits that encode amino acids, known as exons, and then splice them together to produce the final mRNA version of the gene. Human genes are especially rich in introns, with each gene having an average of eight long intronic stretches, so the splicing machinery needs a way to tell where each exon starts and ends.

Research over the past few years has revealed that exons not only specify amino acids, they also contain within their sequences cues necessary for intron removal. Chief among these are exonic splicing enhancer (ESE) motifs—short sequences of about three to eight nucleotides that sit near the ends of the exons and define the exon for the cellular splicing machinery. The need for such motifs can in fact explain a preference for certain nucleotides in human genes. Although the codons GGA and GGG, which encode glycine, can both occur in splicing enhancers, the former codon acts as a more potent enhancer, leading to more efficient splicing. GGA is also correspondingly more common close to the ends of exons.

In support of the view that preserving codon sequence in splicing enhancers matters, research we did with our former University of Bath colleague Joanna L. Parmley has shown that exonic motifs that apparently function as splicing enhancers show slower evolution in their synonymous codons than do neighboring sequences uninvolved in splicing. This slow evolution indicates that natural selection has kept enhancer motifs relatively unchanged because their specific sequences are so significant. Silent alterations to codons containing these enhancers, although they do not change an amino acid, can nonetheless have a major effect on a protein simply because they disrupt the proper removal of introns.

Indeed, when William Fairbrother, now at Brown University, and his colleagues in Christopher Burge's laboratory at the Massachusetts Institute of Technology compared the ends of exons, they found that people are rather similar to one another. These splice-associated regions lack much variation, even at sites where a mutation would be silent. The reason is not that mutations are not happening at the ends of our exons, but that when they appear, the mutations are so damaging to protein production that they tend to be disruptive and disappear from the population.

To date, some 50 genetic disorders have been linked to silent mutations, many of which also appear to interfere with intron removal. Splicing enhancers can overlap with a considerable length of a gene's protein-coding sequence, imposing significant limitations on where a silent mutation would be tolerated. A striking example of the damage a mutation in a splicing enhancer can do was recently documen-ted by Francisco Baralle of the International Center for Genetic Engineering and Biotechnology in Trieste, Italy. The investigators found that 25 percent of the silent mutations they induced in one exon of the *cystic fibrosis transmembrane-conductance regulator (CFTR)* gene disrupted splicing and presumably would thus contribute to cystic fibrosis or related disorders.

That is not to say that disruption of splicing is the only mechanism by which silent mutations can cause disease. Even if the introns are correctly removed from an RNA transcript, the mRNA might not fold properly. Contrary to what is sometimes (for simplicity) depicted in textbooks, an mRNA is not just an unstructured linear string. Like the nucleotide pairs that form between the two DNA strands, separate regions of an mRNA can be complementary and will pair up to create an intricate folded structure known as a stem-loop. The way an mRNA folds determines its stability, which in turn can affect the speed of its translation by ribosomes as well as its subsequent degradation by cellular garbage disposals.

In the *dopamine receptor D2* gene, which encodes a cell-surface receptor that detects the presence of the neurotrans-mitter dopamine, one silent mutation causes the mRNA to be degraded more rapidly than normal. As a result, less of the encoded protein is made, leading to cognitive disorders. Conversely, in the *catechol-O-melhyltransferase (COMT)* gene, a silent mutation increases the extent of mRNA folding, possibly creating too much structure that may be hard to unpack before translation—lowering protein synthesis. Andrea G. Nackley and her colleagues at the University of North Carolina at Chapel Hill found that this mutation affects pain tolerance; perhaps it is no surprise that this research was done in a dental school.

Another instance of a silent mutation affecting a protein is also among the most direct effects and involves a gene known as *multidrug resistance 1.* The gene is so named because its protein product is a cellular pump that in cancer cells helps to expel chemotherapy drugs, thus conferring drug resistance on those cells. Chava Kimchi-Sarfaty and her colleagues at the National Cancer Institute found that the silent change caused the pump protein to misfold, reducing cells' ability to eject drugs. Because the translation process and protein folding can occur simultaneously, the researchers theorized that the rarer synonymous codon produced by the silent mutation caused a pause during translation, which in turn allowed the protein time to adopt an unusual structure. If this pause occurs, the precise cause is unclear and can be added to the list of as yet unsolved mysteries about the workings of genes and proteins.

EFFICIENT GENES, EFFECTIVE MEDICINE

One lesson scientists can take from the recent discoveries about the effects of silent mutations is to be careful in our assumptions. Confidence that synonymous mutations must be "silent" was widespread when there was no mechanism to connect a silent change with an alteration in protein production. But in light of the striking examples described above, this position is no longer tenable.

Recognizing the power of not so silent mutations is beginning to help investigators improve methods for genetic engineering. Knowing which nucleotides in a gene need to be retained and which could potentially be replaced has an immediate application in biotechnology. Both gene therapy and the industrial manufacture of proteins (such as therapeutic drugs), using animals or microbes, rely on the ability to design and fabricate a gene and insert it into a cell's genome. Creating genes that work efficiently is fraught with difficulties, among them ensuring that the newly introduced transgene is acti-vated by the cell, so that adequate amounts of its encoded protein are produced. This is where sensitivity to the effects of synonymous, but not silent, mutations comes in.

In human genes, most introns seem to be dispensable (only one, usually the first intron, appears to be required for the gene to give rise to a protein). This observation means that transgenes can be made compact by removing irrtrons. It also implies that some silent mutation sites could be tweaked without detrimental effect, because leaving out introns does away with the need for splicing enhancers. Released from that constraint, geneticists could exploit those silent sites for other purposes.

A recent experiment at the International Institute of Molecular and Cell Biology in Warsaw illustrates how silent mutation sites could be manipulated for human benefit. Grzegorz Kudla and his colleagues took three genes and did nothing more than change the relative proportion of specific nucleotide bases at the silent sites, then transferred the altered genes to mammalian cells. Remarkably, the investigators found that increasing G and C content led to gene activity and protein manufacture that was up to 100-fold more efficient as compared with GC-poor versions of the same genes.

The new understanding should also inform efforts to understand the root causes of disease. Central to many hunts for the genes underlying diseases are ongoing genome mapping projects to catalogue genetic variation among humans. By identifying all the point mutations, or single-nucleotide polymorphisms (SNPs), in individuals with a given disease, scientists can now home in on regions of the genome containing gene variants that may cause the disease.

Until recently, such searches assumed that if several mutations in a gene correlate with the presence of a disease, those that change the protein's amino acid sequence must be the causal variants. Indeed, the *COMT gene* associated with pain tolerance is known to have a mutation that changes one amino acid to another, and that variant was long assumed to be the only cause of intolerance to pain. Yet individuals with very high or very low pain intolerance were shown to have the same nucleotides at the site of that mutation, indicating there must be some other reason for their differences. Experimental results revealing the silent mutations in *COMT* as the culprits were accepted only once the mechanism by which they might have an effect had been worked out.

It is likely that the causes of some diseases have been wrongly attributed to mutations that change proteins, when in fact synonymous mutations are at fault. Investigators need to keep this possibility in mind as they look for a damaging mutational needle in the genomic haystack. And who knows what additional mechanisms of disease such surveys will turn up?

Do the findings so far mean that silent mutations cause disease frequently? Perhaps. One could argue that silent changes are intellectually interesting and potentially of practical importance, but they are not obviously a cause of much illness. Recent examinations of how genes evolve suggest, however, that such a view may be too complacent. Several years ago one of us (Hurst) showed that in one segment of the *BRCA1* gene (associated with early-onset breast cancer), silent sites evolve very slowly in both rodents and humans. In contrast, the other sequences in this gene segment evolve at a normal rate. The difference does not mean that mutations in the slow-evolving silent sites were rare but rather that individuals who carried the mutations died without passing them on. The segment was later shown to coincide with the location of a splicing enhancer—in other words, it was another example of an area where silent mutation was so damaging its carriers died out.

Just how common are gene segments in which natural selection exerts pressure for the silent sites not to change? Hurst and Parmley investigated that question by scanning genes, searching for regions in which the rate of evolution at silent sites differed markedly from the rates at sites that change amino acids. To our surprise, we discovered that DNA segments containing unusually slow-evolving silent sites are relatively common. Indeed, they are more prevalent than regions where encoded amino acids evolve exceptionally quickly. A stretch of very highly conserved silent mutation sites occurs, on average, once every 10,000 to 15,000 nucleo-tides of gene sequence.

We estimate that between 5 and 10 percent of human genes contain at least one region where silent mutations could be harmful. Peter Schattner and Mark Diekhans of the University of California, Santa Cruz, performed a similar analysis, looking for large regions in genes with unusually hyperconserved silent sites. They estimated that there were about 1,600 such blocks in the nearly 12,000 genes they examined, which approximately corresponds with our estimates. Both these estimates are, however, likely to be low, and the true figure may well be considerably greater. If these conserved sites indicate the locations of silent mutations that can cause disease, as seems probable, ignoring them will inevitably lead to misidentification of disease-associated mutations.

Recognition that natural selection does take notice of not so silent mutations has gone hand in hand with scientists' realization that the process through which genes make proteins is vastly more complex and more nuanced than previously imagined. The way genes evolve and the ways in which they work are also intimately coupled to a degree barely understood even a decade ago. Further study of both these processes will continue lo illuminate the remarkable complexity of the workings of genomes. DNA, for example, is not simply a linear molecule but is wrapped into coils and must be unpacked to enable transcription. Does the control of this process leave a fingerprint on silent sites as well? Likewise, runs of rare codons are more common than they should be, but what are they doing and why?

Answers to these and related questions will not just make greater sense of making proteins, they could well provide insights that cure disease.

BIASED VACCINE

Manipulating the sites of synonymous mutations has allowed scientists to design genes that speed up protein manufacture, but the same technique can also be used to slow it down. Steffen Mueller and his colleagues at Stony Brook University recently took this approach to design a safer polio vaccine. Live viruses make the most potent vaccines because they provoke a strong immune response in the recipient, but live vaccines can reproduce and mutate, potentially causing disease. Mueller's group took advantage of microbes' preference for using certain codons to maximize the efficiency of protein production by designing a poliovirus that substituted rarer, less efficient codons in sequences encoding the viral shell. The resulting virus was able to copy itself, albeit more slowly. After the investigators administered the engineered virus to mice, the animals were protected from infection when they were later exposed to wild poliovirus. This technique for taking advantage of codon bias to create a live but weakened vaccine could be applied to other pathogens as well to produce potent but safer vaccines.

3. LAURIE GARRETT, THE PERSON

Laurie Garrett (born 1951) is an American science journalist and author. She was awarded the Pulitzer Prize for Explanatory Journalism in 1996 for a series of works published in *Newsday*, chronicling the Ebola virus outbreak in Zaire.

BIOGRAPHY

Laurie Garrett was born in Los Angeles, California, in 1951. She graduated from San Marino High School in 1969. She then graduated with honors from Merrill College at the University of California, Santa Cruz, where she received a B.S. in biology in 1975. She attended graduate school in the Department of Bacteriology and Immunology at University of California, Berkeley and did research at Stanford University with Leonard Herzenberg.

During her PhD studies, Garrett started reporting on science news for radio station KPFA. The hobby soon became far more interesting than graduate school and she took a leave of absence to explore journalism, and has never completed that degree. At KPFA, she worked in management, in news, and in radio documentary production. A documentary series she co-produced (with Adi Gevins) won the 1977 Peabody Award in Broadcasting, and other KPFA production efforts by Garrett, won the Edwin Howard Armstrong award.

In 1997, Garrett won a George Polk Award for Foreign Reporting, for "Crumbled Empire, Shattered Health" in *Newsday*, described as "a series of 25 articles on the public health crisis in the former Soviet Union". She won another Polk award in 2000 for her book *Betrayal of Trust*, "a meticulously researched account of health catastrophes occurring in different places simultaneously and amounting to a disaster of global proportions".

In 2004, Garrett joined the Council on Foreign Relations, as the Senior Fellow of the Global Health Program. She has worked on a broad variety of public health issues including SARS, avian flu, tuberculosis, malaria, shipping container clinics, and the intersection of HIV/AIDS and national security. Garrett lives in Brooklyn Heights, New York City.

PUBLISHED WORKS

✓ Books:

- *I Heard the Sirens Scream: How Americans Responded to the 9/11 and Anthrax Attacks. Amazon.com Kindle e-book. 2011.*
- *Betrayal of Trust: The Collapse of Global Public Health. Hyperion. 2001. ISBN 0-7868-8440-1.*
- *The Coming Plague: Newly Emerging Diseases in a World Out of Balance. Penguin. 1995. ISBN 0-14-025091-3.*
  ... which discusses the vulnerability of the world to disease, due to the lack of attention, and of funding, given to health

✓ Articles:

- (2020). Trump Has Sabotaged America's Coronavirus Response. *Foreign Policy*
- (2015). Ebola's Lessons, How the WHO Mishandled the Crisis. *Foreign Affairs*
- (2005). The Next Pandemic. *Foreign Affairs*
- (2001). The Nightmare of Bioterrorism. *Foreign Affairs*

THE AUTHOR

Laurie Garrett wrote her first bestselling book, *THE COMING PLAGUE: Newly Emerging Diseases in a World Out of Balance*, while splitting her time between the Harvard School of Public Health and the New York newspaper, *Newsday*. In the 1992-93 academic years Garrett was a Fellow at Harvard, where she worked closely with the emerging diseases group, a collection of faculty concerned about the surge in epidemics of previously unknown or rare viruses and bacteria. The book was published in hardcopy by Farrar, Straus, and Giroux in 1994, and spent 19 weeks on the *New York Times* bestseller list. Released in paperback in 1995 by Viking Penguin, *THE COMING PLAGUE* remains in print and continues to sell vigorously. The book was released in e-book form in May 2011.

During the 1990s Garrett continued tracking outbreaks and epidemics worldwide, noting the insufficient responses from global public health institutions in Zaire, India, Russia and most of the former USSR, Eastern Europe, and the United States. This resulted in publication in 2000 of *BETRAYAL OF TRUST: The Collapse of Global Public Health*, released by Hyperion that year, and in paperback in 2001. *BETRAYAL OF TRUST* was also a vigorous seller, and remains in print today. It was also released as an e-book in May 2011.

In Summer 2011 , Garrett's long-awaited third book, *I HEARD THE SIRENS SCREAM: How Americans Responded to the 9/11 and Anthrax Attacks*, was published in time for the 10th Anniversary of 9/11 by Amazon as an e-book. It is also available on Kindle and in paperback.

Garrett is a Member of the World Economic Forum Global Health Security Advisory Board. She is also a member of the Council on Foreign Relations, and the National Association for Science Writers.

THE CFR

In 2004, Laurie Garrett left *Newsday* to join the think tank staff of the Council on Foreign Relations in New York. She created the Council on Foreign Relation's Global Health Program, where she served as Senior Fellow for Global Health from 2004-2017. Garrett has written several reports and articles including: *HIV and National Security: Where are the Links?, A Council Report* (Council on Foreign Relations Press, 2005), "The Next Pandemic?" (Foreign Affairs, July/ August 2005), "The Lessons of HIV/AIDS" (Foreign Affairs, July/ August 2005), "The Challenge of Global Health" (Foreign Affairs, January/February 2007), *The Future of Foreign Assistance Amid Global Economic and Financial Crisis, A Council on Foreign Relations Action Plan* (2009), and *CastroCare in Crisis* (Foreign Affairs July/August 2010). Learn more about Laurie's work at CFR on their website.

THE "PUBLIC INTELLECTUAL"

Dr. Declan Fahy of Dublin City University presented the following abstract at the Public Communication of Science and Technology conference in Istanbul in April 2016 — "Knowledge-based reporting of global infectious diseases: The pioneering journalism of Laurie Garrett".

"I explain in this paper how Pulitzer Prize-winning American health journalist Laurie Garrett became one of the world's most influential and authoritative writers on global infectious diseases.

I take Garrett as a case study of what journalism scholar Thomas Patterson and communication researcher Wolfgang Donsbach have called knowledge-based journalism. That is, journalism that applies one or more fields of specialized knowledge to the coverage of complex events, such as pandemics, in order to enhance citizens' understanding and influence policy on these new social threats.

To demonstrate how Garrett came to undertake this form of journalism, I examine the key moments of her career from the 1980s, when she first reported on the new disease of AIDS for the newspaper Newsday, to her most recent journalism on the 2014 Ebola outbreak in her role as Fellow at the US think tank, the Council for Foreign Relations.

Using methodological approaches from cultural and intellectual history, I analyze chronologically her news reports, books - The Coming Plague (1995), Betrayal of Trust (2000) and I Heard the Sirens Scream (2011) - policy reports and social media activity to show how she came to offer a unique, authoritative perspective on global infectious disease. I also situate these texts against the a historical background, where the new field of emerging infectious diseases (EID) developed in the 1990s as a new scientific framework for understanding novel threats such as HIV/AIDS, SARS, bird flu, and Ebola.

4. U.S. - CHINESE DISTRUST IS INVITING DANGEROUS CORONAVIRUS CONSPIRACY THEORIES

And Undermining Efforts to Contain the Epidemic

By Yanzhong Huang, March 5, 2020

Coronavirus researchers at the Enshi Center for Disease Control and Prevention in China, February 2020 Yang Shunpi Xinhua / eyevine / Redux

Tens of thousands of people get sick. More than 2,900 die. Fear spreads faster than the virus. Factories are closed. Roads are blocked. Villages are sealed off. Cities are locked down. The outbreak of the 2019 novel coronavirus (COVID-19) is the most severe sociopolitical crisis Chinese leaders have grappled with since the 1989 Tiananmen crackdown. And the crisis is not confined to China. The spread of the virus across borders—and the panicky reaction to that spread—will have profound effects on the global economy, politics, security, and governance.

The virus's novelty leaves many unknowns. We still don't have a clear idea of its transmissibility and virulence. We do not have a clear idea of the incubation period, which could last up to 24 days. We also don't know how infectious people are before their symptoms manifest and why some cases suddenly become severe. We also don't understand why some patients tested positive a second time even after they seemingly recovered.

Rumors thrive on fear and uncertainty, and the outbreak of the novel coronavirus offers plenty of both. Within weeks of the pathogen's appearance, social media lit up with suggestions that the virus was a biological weapon—either a Chinese one that had escaped from a laboratory in Wuhan or an American one inflicted on Wuhan. While such rumors are not credible, given that neither the United States nor China has incentive to develop biological weapons, they are difficult to dispel, because military officials on both sides still view with suspicion each other's motives in building biosecurity programs. Loopholes in China's biosafety regulations only allow the rumors to gain more currency. And the lack of trust between the two nations—as evidenced by China's initial refusal to allow U.S. disease experts to visit Wuhan—is undermining efforts to contain the virus's global spread.

THE THEORY

From the beginning of the outbreak, the public was informed that the disease originated with human exposure to a virus carried by wild animals. But the suddenness and mystery of the bug's appearance left fertile ground for speculation, and soon enough, online sources began to advance the claim that the virus was genetically engineered. An unpublished paper authored by Indian scientists seemed to bolster this notion by suggesting that the virus's protein sequence included elements of HIV, the virus that causes AIDS. Although the authors soon voluntarily withdrew the paper, the proposed linkage caught the attention of websites such as Zero Hedge, which claimed that the novel coronavirus was weaponized by Chinese scientists. Speaking on Fox News, Tom Cotton, the Republican senator from Arkansas, suggested that it could not be ruled out that the virus originated in a lab in Wuhan that is used to handle the most dangerous pathogens.

Zero Hedge has been barred from Twitter, but Chinese social media abounds with conjecture that the virus was engineered by the United States as an agent of biological warfare against China. One widely shared conspiracy theory suggests that American soldiers participating in the 2019 Military World Games in Wuhan deliberately shed the virus at the Hunan Seafood Market. Contending that "a new type of biological warfare is coming," a retired People's

Liberation Army general called for building a permanent biodefense force in China.

The suddenness and mystery of the bug's appearance left fertile ground for speculation.

The current outbreak in China is not the first to be a rumored biological weapons attack. During the 2002–3 SARS epidemic, a Russian scientist claimed that the virus was a mixture of measles and mumps that could be made only in the lab. Many Chinese seized on this notion and speculated that SARS was a genetic weapon developed by the United States to target them alone. The official *China Youth Daily* linked a National Institutes of Health–sponsored genetic study in China to the U.S. genetic warfare program. In the United States, meanwhile, a China expert suggested that the virus was linked to China's biowarfare program. Yet SARS was by no means a genetic weapon. According to the U.S. Centers for Disease Control and Prevention, of the 166 reported SARS patients in the United States in 2003, 58 percent were white and 32 percent were Asian.

THE HISTORY

Just how likely is either the United States or China to be developing deadly biological weapons for use? A tour through the history of such warfare is instructive.

During World War II, the United States developed biological weapons but never used them. Biological agents had certain liabilities for battlefield use: they didn't take effect right away, they could infect one's own forces, they were sensitive to environmental and meteorological conditions, and they could conceivably contaminate an area for longer than intended. Nonetheless, the United States continued to stockpile and develop biological weapons into the postwar era.

Matthew Meselson, a biologist at Harvard University, led a successful campaign against biological weapons development starting in the early 1960s. In 1969, the United States got rid of its offensive biological warfare program and played a crucial role in successfully negotiating an international treaty known as the Biological Weapons Convention (BWC). The treaty prohibits the development, production, and stockpiling of biological agents and related delivery systems intended for hostile use. In explaining the U.S. decision, President Richard Nixon commented in 1970 that "we'll never use the damn germs, so what good is biological warfare as a deterrent? If somebody uses germs on us, we'll nuke 'em."

Compared with the United States, China came late to the game. The country had been on the receiving end of germ warfare, on the part of the Imperial Japanese Army's biowarfare Unit 731 during World War II. As a result, China felt an imperative to build research facilities devoted to "defensive" biological warfare. In August 1951, Premier Zhou Enlai set up the Academy of Military Medical Sciences (AMMS) to conduct research on biodefense against "wartime special weapons."

PARAMILITARY IN FRONT OF A MONUMENT OF ENLAI IN JIANGSU PROVINCE, CHINA, MARCH 2008 CHINA DAILY CDIC / REUTERS

Since China did not possess nuclear weapons until the mid-1960s, it may indeed have explored developing biological weapons as a weapon of last resort or a strategic deterrent similar to nuclear weapons. But by 1982, China had acquired a largely invulnerable retaliatory nuclear arsenal. Two years later, China acceded to the BWC. The timing indicates that China, like the United States, found nuclear weapons to be the more credible and effective deterrent.

The mid-1980s saw a shift in China's national agenda toward economic development. Funding for China's biodefense research facilities dwindled, and they began developing products for civilian rather than military purposes. The AMMS became something of an analog to the U.S. Army Medical Research Institute of Infectious Diseases. It developed a pan-antimalaria drug called compound benflumentol and registered patents in more than 50 countries. During the 2014 Ebola virus outbreak, AMMS collaborated with Chinese pharmaceutical companies to develop two drugs for treating the deadly disease.

DISTRUST AND MISPERCEPTIONS

China and the United States are both parties to the BWC, but they still look upon each other with suspicion. Past U.S. government reports have alleged that China continued to possess "an offensive biological warfare capability based on technology developed prior to its accession to the BWC." According to a former official from the U.S. Department of Defense, by the 1990s China had manufactured and weaponized a wide variety of infectious microorganisms and toxins and had a wide range of delivery means available, including ballistic and cruise missiles. Although these reports and accusations have never been substantiated by open-source evidence, official Chinese publications do suggest sustained and organized biowarfare-related research activity. Official sources reported that in the 1990s, Chinese scientists used rare earth as a medium in which to swiftly cultivate brucellosis (traditionally considered a biological agent suitable for military use).

Many people in China also perceive the United States as a potential biological warfare threat. After the 2002–3 SARS outbreak, some Chinese military experts invoked a scenario of enemies spraying unknown SARS-like viruses on Beijing during airstrikes. Noting that the United States had developed antibiotic-resistant anthrax strains, a leading Chinese military medical expert implied that Washington had weaponized SARS and avian flu virus. In 2001, the administration of President George W. Bush had rejected a proposed protocol to the BWC on the grounds that it was insufficient to its purpose. That refusal convinced some Chinese experts that the United Sates had renewed its interest in developing biological weapons. In 2007, Chinese military researchers published an article accusing the United States of "using new technologies to develop novel biological weapons agents" and claiming that it was "extremely likely" that anthrax spores in the 2001 attacks on Democratic senators' offices came from U.S. military labs. Such suspicions might explain why the Chinese government later tightened regulations on foreigners using human genetic material and made it more difficult to pass the material abroad.

The mutual distrust and misperceptions are emblematic of a classic security dilemma, in which actions taken by one state to improve its security lead to reactions from others, which make the original state less secure. Worse, biodefense programs are so opaque, and provoke such moral antipathy, that they encourage "looking-glass presumptions": when one state is perceived to be pursuing biological weapons, its rivals will likely seek to acquire them as well.

During World War II, the United States, Canada, and the United Kingdom all developed biological weapons because they thought Hitler's Germany would develop them (it didn't).

Many in China also perceive the United States as a potential biological warfare threat.

Whether a disease is naturally occurring or deliberately caused can be difficult to tell, because many biological agents are naturally accessible, and their production is of dual use up to the point of weaponization. In the context of frigid bilateral relations, a naturally occurring disease outbreak caused by an unknown pathogen can be easily framed as a bioweapons attack. The historian Alfred Crosby noted that the 1918 Spanish flu pandemic was suspected to have been started by German agents. In 2004, the Indian government accused "promiscuous Pakistanis" of conducting Islamic "jihad terrorism" by deliberately spreading HIV in Kashmir. When H5N1 ("bird flu") became a major concern worldwide in 2008, Indonesia's then-health minister, Siti Supari, accused the United States of using virus samples to develop biological weapons and suspended the operation of a U.S. Navy medical research unit in Jakarta.

At a time of deteriorating relations between the United States and China, misperceptions of a hostile origin of COVID-19 have undermined global efforts to tackle the pathogen's spread. For weeks, China ignored offers of help from the U.S. Centers for Disease Control and Prevention. According to a blog post published on a website affiliated with *Jiefang Daily* (the official newspaper of the Shanghai Committee of the Communist Party of China), "Some U.S. CDC experts" might be on a military mission to "spy on China's virologic research capacity." Two U.S. experts finally joined the World Health Organization's delegation to China in February, but the delegation's field visit did not include the Wuhan Institute of Virology—indeed, Wuhan, the epicenter of the outbreak, was initially not even on the group's itinerary.

The claim that the novel coronavirus is a biological weapon is not only harmful but also scientifically unsupported. Scientists have pointed out that mutations in the virus are "completely consistent with natural evolution." According to *The Lancet*, scientists from multiple countries have "overwhelmingly" concluded that the novel coronavirus originated in wildlife.

AN ACCIDENTAL RELEASE?

Since coronaviruses are zoonotic, which is to say that they jump from animals to humans, scientists believe that they have an animal origin. Most agree that bats are the natural reservoir host for the virus, although snakes for a time were also suspected. Scientists conjecture that pangolins—widely considered a delicacy in China—may be the intermediate host to pass the novel virus to humans. Overall, they are inclined to view the outbreak as a problem of zoonotic infection transmitted from wild animals to humans. While the virus was indeed found in people associated with a wet market (where live animals were sold and slaughtered) as well as in the market environment, some of the early cases were in people who had not visited the market, suggesting that the jumping of species may have occurred elsewhere or earlier.

Another, less supported hypothesis views the outbreak as the result of a biosafety accident in which the coronavirus leaked out from a lab where scientists failed to follow proper decontamination protocol. Supporters of this theory point to the seemingly overwhelming circumstantial evidence that connects the outbreak to the Wuhan Institute of Virology, which also houses China's only biosafety level 4 (BSL-4) lab, the highest level of biosafety precautions. They noted that Dr. Shi Zhengli, a researcher in the lab who has acquired the nickname "Batwoman," has been actively hunting down the coronavirus and proving that bats are natural reservoirs for SARS-like coronaviruses. Shi has flatly denied that the institute was the source of the novel coronavirus, which she said was rather "nature punishing the human race for keeping uncivilized living habits."

Other Chinese social media posts focused on the credentials of the director general of the institute, Wang Yanyi. A widely shared post allegedly written by Rao Yi, a leading Chinese biologist, said Wang had a weak academic background and ascended to her current post through nepotism. On February 17, one post accused Wang of flouting the biosafety rules and selling laboratory animals to wet markets for profit. Another study, conducted by the South China University of Technology, concluded that the coronavirus "probably" originated in the Wuhan Center for Disease Control and Prevention, which is just 280 meters away from the Hunan Seafood Market.

The study mentioned that bats linked to coronavirus once attacked a researcher, who had to be self-quarantined because "blood of a bat shot on his skin." The paper was later removed from ResearchGate, a commercial social-networking site for scientists and researchers to share papers. Thus far, no scientists have confirmed or refuted the paper's findings.

THE CLOSED SEAFOOD MARKET IN WUHAN, JANUARY 2020 DARLEY SHEN / REUTERS

Many of these accusations may be groundless, but nobody can deny that lab safety is a major concern in China. A safety breach at a Chinese Center for Disease Control and Prevention lab is believed to have caused four suspected SARS cases, including one death, in Beijing in 2004. A similar accident caused 65 lab workers of Lanzhou Veterinary Research Institute to be infected with brucellosis in December 2019. In January 2020, a renowned Chinese scientist, Li Ning, was sentenced to 12 years in prison for selling experimental animals to local markets.

The government's actions have lent credibility to the thesis that the coronavirus accidentally escaped a laboratory. In February, China appointed Major General Chen Wei, China's top biowarfare expert, as head of the BSL-4 laboratory at Wuhan Institute of Virology. Because of Chen's background, the appointment fueled suspicions about the virus's possible connection to the BSL-4 lab. Then, on February 14, Chinese President Xi Jinping highlighted the need to incorporate *shengwu anquan* (which in Chinese could mean either "biosecurity" or "biosafety") into its national security regime.

In the literature of biological warfare, the difference between biosecurity and biosafety is important: the former is about the protection of humans and the environment from the *intentional* release of pathogens and biohazards, while the latter is about safety from their *unintentional* release. Xi's remarks were immediately followed by a Ministry of Science and Technology instruction on strengthening biosafety management in labs handling the novel coronavirus, suggesting that Xi had biosafety in mind when issuing the directive. But the Chinese leader could as easily have been referring to animal agriculture, where biosecurity is broadly defined as everything done to keep disease away from animals and the people that may interact with them. Indeed, just last week, China's legislature announced a permanent ban on wildlife trade and consumption in the country, apparently to minimize the chances of diseases passing to humans from animals.

AN IMPEDIMENT TO COLLABORATION

The virus is now rapidly spreading worldwide, with social, political, and economic consequences wherever it goes. Identifying its origin would help experts and governments to hone the best countermeasures to stem its spread and prevent such outbreaks in the future.

So far, neither the theory that the virus was developed as a biological weapon nor the notion that it escaped a laboratory by accident seems as plausible as the hypothesis that the virus jumped to humans from animals at the wet market. But the conspiracy theories have poisoned the atmosphere for U.S.-Chinese collaboration in addressing the outbreak, which might otherwise have presented an opportunity to reset the soured relationship.

In order to dispel misperceptions and minimalize the damage to future relations, the two countries should consider expanding their military-to-military exchanges, such that they might visit each other's sites for conducting government-sponsored biodefense work. And the United States should explore channels for helping China improve its laboratory biosafety. The beginning of either measure is dialogue.

INDEX

B

C

D

F

G

H

I

N

O

P

T

U

V

W